Global Eduction Tectonics of the Expanding Earth

Global Eduction Tectonics of the Expanding Earth

Yu.V. Chudinov
Doctor of Geology,
Seismological Centre, RAS

///VSP///
Utrecht, The Netherlands, 1998

VSP BV
P.O. Box 346
3700 AH Zeist
The Netherlands

Tel: +31 30 692 5790
Fax: +31 30 693 2081
E-mail: vsppub@compuserve.com
Home Page: http://www.vsppub.com

First published in 1998

ISBN 90-6764-280-0

Printed in The Netherlands by Ridderprint bv, Ridderkerk.

CONTENTS

FOREWORD

This book is a new approach in the development of global tectonics.

Over many years, geologists and geophysicists have tried to find the answer to the question: why is the shape of continents so similar that they could be gathered in one single precontinent. Several hypotheses of the Earth's evolution were influenced by this phenomenon. Created by A. Wegener the hypothesis of continental drift, popular in the 1910s-1920s, was forgotten for some time, only to appear again in the form of new global plate tectonics after a revolutionary discovery of ocean floor spreading in the early 1960s.

An alternative to the Wegener hypothesis — the hypothesis of the Earth's expansion which explained the mutual removal of the continents by expansion of the ocean floor surface — was developed from the late 1920s by O. Hilgenberg and some other researchers, but these ideas were less popular than the notion of continental drift, and less so also than the recent plate tectonics concept. However, in essence, it had explained the cause of ocean spreading before the latter was revealed.

Both hypotheses — plate tectonics and the Earth's expansion — adopt the fact of the spreading of the ocean floor, but they differ greatly in explaining the processes of new floor origination, movement of continents, and mantle convection. Nevertheless, their adherents had no real discussion. The debates, if they existed, were held, in the Russian terminology, between mobilism and fixism, and their main viewpoint was either that the continents are moving apart while the oceans are expanding, or not. It is like a choice between white and black. But there are, besides, many-coloured variants, with the Earth's expansion concept among them.

Being accepted and supported by the majority of scientists, plate tectonics won the competition and led in the course of the last 25 years. As commonly occurs, not all opponents laid down their arms however, and while plate tectonics was broadly developed by the geological community, apologists of the Earth's expansion theory continued the accumulation of new geological, geophysical and geochemical evidence. It looked as though new evidence was getting more and more difficult to accommodate by plate tectonics. At the same time it seemed to fit more naturally into the Earth's expansion hypothesis.

In the case under consideration, the cornerstones of a new variant of the Earth expansion concept became the model, originated by the author of this book, of `eduction', i. e. of lifting up the mantle material onto the Earth's surface in the active margins of oceans and non-closed character of the mantle currents.

Yury Chudinov is one of the brightest and most successful supporters of the Earth expansion ideas. He developed the deepest and most comprehensive scientific scenario of tectonic evolution of the Earth using the idea of its radius increase. In 1979 he first published short reports on the understanding of the main tectonic processes characteristic of active borders of the continents and oceans by the new eduction theory, and in 1985 the first version of this book appeared. In the considerable part of his studies the author appeals to clearly demonstrated evidence which cannot be disputed. The significant and most interesting part of the book is a critical analysis of the history of development of the plate tectonics, which is done in a masterly way from his specific viewpoint. The most difficult part of the whole concept is the physical explanation of the Earth's radius increase. However, taking the point of view of a geologist, one should accept the Earth's expansion: the absence of its physical explanation seems to be a problem for physicists, but not for geologists.

A remarkable part of the book is the explanation of existing seismicity — its specific and geographical distribution from the point of view of tectonic evolution of the lithosphere. This part of the book has a specific interest for seismologists.

Yu. Chudinov is critically minded, scrupulous by nature, and sees with a broad perspective. Perhaps, analysing facts and considering existing hypotheses he is more indulgent to ideas of expansion. But he proceeds mainly from the facts, and the amount of speculation is very limited. I foresee some scepticism of apologists of plate tectonics about Yury Chudinov's notions, but I believe that this book should be a strong trigger for the critical review of contemporary tectonic concepts that they really need.

Alexey V. Nikolaev,

Professor of geophysics,
President of Seismological
Association of Russia

PREFACE

This book is a reworked edition of the monograph entitled `Geology of Active Oceanic Margins and Global Tectonics' issued by the Moscow Nedra publishers in 1985: the new title reflects more accurately its contents.

The author began his geological research in 1951. Since that time, and up to 1988, it was accompanied by annual field work in different regions of the former USSR and Mongolia. Within this large territory (and even within the territory of present Russia too) one can become familiar with virtually all types of geological structures and different-age formations to be found on the Earth.

A variety of geological circumstances demanded that the author should gather together and present his observations and theories. In deriving them, the author relied first of all on direct geological observations, both his own and those of other researchers, since the current literature has become diffuse and contradictory. The main value of the literature available, in the author's opinion, is its ability to acquaint us with facts that were previously unknown and as a source of fruitful ideas. So, the author's views have been mostly influenced by works of such Russian tectonists as A. V. Peive, S. S. Kuznetsov, Yu. M. Sheinmann, V. M. Sinitsin, and others.

The most fruitful ground for material for this book was the author's work in the North-East of Russia — a large region within which the author has spent most time in field investigations (1973-1988). The beginning of this work coincided with the dissemination of new ideas of global tectonics: the information obtained by the investigations of the author and other researchers of the Russian North-East proved to be intimately bound up with questions raised by plate tectonics, especially with those related to the geological history of the Pacific active margins. The solutions to these questions suggested in plate tectonics did not satisfy the author; moreover, after several years work in the region, the author arrived at the conclusion that not only was the subduction model unreliable, but that on the active margins, there occur quite the opposite tectonic movements which were called by the author `eduction'. At first, this conclusion was quite unexpected even for the author himself, but later he became convinced that it was in perfect agreement with numerous facts.

The reader may notice that the literature list contains a considerable number of works from the 1960s and 1970s; this is largely due to the necessity of presenting a treatment of plate tectonics history. The first steps of its origin are now fading into oblivion, and comprehension of its hypothetical character and its inaccuracies is being replaced by the belief that in those early days, proponents of the theory had thought everything out and made correct and definite decisions on fundamental questions. However, one cannot ignore the fundamentals of a theory in an attempt to develop its details as soon as this seems to be possible. From my point of view, a feature in the development of plate tectonics in recent years has been its departure from fundamental questions and concentrating instead on several particular ones. This makes it essential to consider the basis of fundamental assumptions and deductions in the early evolution of the concept.

It is worth mentioning that in recent years geological explorations in Russia have been substantially curtailed, so that new geological data from northeastern Russia are relatively scarce.

The author is grateful to the translators of the book — O. Meshkov and V. Agranat (Chapter VI) — who did some linguistic editing of the English text written by the author in addition to the translation proper. The author also thanks N. Vladimirskaya and A. Artemov whose invaluable help contributed greatly to the preparation of this book. However, any obscurities in presentation that remain are the responsibility of the author.

INTRODUCTION

It is the author's belief that proper comprehension of geological processes occurring in the Pacific, as well as over the Earth as a whole, depends mainly on the determination of the direction of movements which takes place along the deep seismofocal (Benioff-Zavaritsky) zones situated on the oceanic margins. Recently, a hypothesis has been advanced suggesting that these zones are characterized by a thrusting of plates under one another by subduction, as it is commonly termed. However, some researchers are of different opinion, maintaining that there are no unidirectional movements along these zones. They are considered instead to be an area of relaxation of regional stresses, with a disorderly orientation of individual displacements. Also, the concept that seismofocal zones are conductors of energy from depth toward the Earth's surface has been proposed (with a new name, `tectonoferas' for such zones). However the author's point of view is that the prevailing movements in the seismofocal zones are the out-thrusting of their lower wings (those that face the ocean) from the depths to the Earth's surface. The process involving movements in this direction is quite the opposite to subduction and is thus termed `eduction' by the author (Latin: educere — to draw out) by analogy with existing terms such as subduction, obduction, etc.

This work aims at providing evidence for the real existence of eduction. The author has made use of the data obtained from geophysical investigations of seismofocal zones, evidence on the structure and history of geological development of North-East Russia and other regions, as well as evidence obtained from deep sea drilling. The book also considers the application of the eduction concept to the main problems of the global tectonic development of the Earth.

The author has also devoted significant attention to a critical analysis of the subduction concept that constitutes an integral part of new global tectonics, or the plate hypothesis (plate tectonics). The essence of this hypothesis is the assumption that ocean expansion caused by the spreading of the mid-oceanic ridges is compensated for by submersion of superfluous lithosphere in subduction, i.e. seismofocal zones; compensation of expansion by subduction keeps the Earth's size constant. The subduction process is as important for plate tectonics as is the idea of sea-floor spreading, but it should be noted that the works by adherents of the plate concept discuss particular problems of subduction much less than those of the spreading phenomenon. Such disparity is explainable: the spreading phenomenon can be proven by various factual geological and geophysical data which are constantly replenished while the subduction one is being substantiated with circumstantial interpretation of geophysical materials and speculative conclusions. Furthermore, the evidence for the spreading process accompanied by factual details is often used as a basic proof for the truth of the plate tectonics concept as a whole. Posing the problem in such a way is undoubtedly wrong; and occasional statements saying that the existence of spreading confirms the existence of subduction are groundless. There is no doubt that the subduction model constitutes the weakest link in the construction of plate tectonics, as has repeatedly been pointed out.

It is worth noting that application of a subduction model to geological data neither contributes any new details to geological constructions, nor does it make geological history more accurate. There are no publications using a subduction basis that treat development of individual regions in a convincing and detailed enough way i.e. comparable to the details available in existing geological maps. The graphic illustration in such works is usually represented with small-scale schemes where the few arbitrary boundaries are not strictly fixed geographically. Now that the subduction concept has been developed for almost 30 years, it can be said that it has not been fruitful geologically.

It seems quite obvious that general comprehension of global processes depends on how the problem of direction of movements along seismofocal zones is solved, using the concepts of either eduction or subduction. If one assumes the existence of subduction, one can consider the Earth's size to be constant or changing but slightly. Eduction leads to the growth of ocean floor from the side of marginal trenches in addition to spreading, so this process is hardly likely unless the Earth's size increases considerably.

A choice of either subduction or eduction in direction of movements also has practical consequences. Occurring above the recent and ancient seismofocal zones are the largest belts of endogenic ore deposits containing the major Earth resources of various metals. The eduction and subduction approaches yield essentially different explanations for the origin of these belts.

The main source for ore material, if viewed subductionally, is submerged oceanic crust and sediments overlapping it which are analogues of recent spreading-originated crust and sediments. It is known that the latter are often devoid of such ore components and only very rarely contain some of the metals typical of the above-seismofocal ore belts, for example, bismuth, mercury or antimony. Besides, assuming subduction, one should admit that the ore-forming and other processes which are due to the tectono-magmatic activization within the above-seismofocal area, would be supplied with a minimum amount of heat energy theoretically necessary. A `starvation ration' of required heat energy stems from the difficulties experienced by the subduction hypothesis when accounting for heat sources.

Eductionally, the source of ore matter is a depth material which starts its movement at the base of the upper mantle and travels gradually upwards until it just escapes onto the Earth's surface. No doubt, this is originally slightly differentiated material capable of containing many components foreign to near-surface parts of the lithosphere. In this case, heat reserves are delivered into the higher parts of the lithosphere, together with matter transported mechanically from a depth of several hundreds of kilometres; such heat considerably exceeds that admitted in traditional geological notions, the more so in the plate tectonics construction. Then, fluid and gas transfer of ore material may be of more importance in ore formation than normally supposed.

Chapter I. Seismofocal Zones

SOME PROBLEMS OF DEPTH GEOLOGY AND TERMS USED

In recent decades, depth geology, which studies the structures not only of the Earth's crust but also of the upper mantle, has been gaining increasingly greater attention. Advances in this branch have resulted in a number of achievements, one of the earliest and most important of them being a concept of deep-seated faults, elaborated by A. V. Peive in several publications from 1945 on. It radically changed the practice of geological explorations and mapping in the Soviet Union.

However, the deep-seated faults concept was treated by some researchers in a rather too one-sided way when the faults were regarded as independent autonomous structures capable of controlling sedimentary, magmatic and other geological processes, with no consideration of the composition and structures of the masses divided by them. This view found expression in tectonic schemes, maps and pictures, where both wings of deep-seated faults were shown as a one-colored field, i.e. without analysis of any differences between them, or even in the form of lines on an unpainted background. This way of conceiving the geological situation in relation to the depth faults is to be found still in many modern plate tectonics works.

But starting from his early works devoted to deep-seated faults, Peive (1945,1947) treated them as the boundaries of depth structures existing both in the crust and in the upper mantle. It was just such a treatment which was used in considering the interrelation of depth faults and depth structures in the region of their initial identification on the eastern slope of the Urals (Peive, 1945). So, in a paper on the development of the Chu-Ili geosyncline in Kazakhstan, Peive wrote: `By studying the geological structure of orogens using only geological methods and making no use of geophysical data, one could, nevertheless, obtain quite definite, although circumstantial, evidence for the participation in tectonic movements of lithospheric masses lying at depths of hundreds of kilometres' (Peive, 1954, p. 107). Current as well as earlier advances of depth geology continue to be defined by revealing the mantle-crust lithospheric structures divided by great deep-seated faults.

At an early stage of development of the deep-seated faults concept, seismofocal zones were considered by A.V. Peive (1946) and N.S. Shatsky (1946) to be a variety of these faults. However, the idea of deep-seated faults was extended to include a much wider series of fault structures to embrace tectonic boundaries of the continental and oceanic crust of a non-active type similar to recent Atlantic borders, suture boundaries resulting from rapprochement of heterogeneous blocks accompanied by destruction and compression of masses which had previously divided them, and just large faults, especially if associated with an intrusive formation of any composition, primarily ultrabasic bodies. But probably, tectonic boundaries corresponding to the modern seismofocal zones of a depth of 700 km, and their paleoanalogues too, must be separated from the general group of depth faults as a special and as the largest variety. It should be emphasized that they are not faults in a common geological sense, i.e. they are not surfaces of tectonic displacements of more or less uniform matter, but such tectonic zones which divide geological masses of always different composition and possibly, in some cases, of different physical state. It is reasonable to believe that such dividing structures might have originated only simultaneously with the general structures of their wings. These tectonic surfaces are the deepest among the tectonic boundaries known to exist on

Earth. Their nature is dealt with throughout almost the whole book, but particularly in this chapter.

The influence on the development of depth geology of what is now the most popular concept of plate tectonics merits some consideration. As a whole, the contribution of the plate tectonics concept to the general theory of geology has had both benefits and drawbacks. On the one hand, it destroyed a number of traditional geological conceptions and showed that they were not so reliable as they had seemed to many geologists. Furthermore, the new concept was created substantially by geophysicists, and their information made available for elaborating geological theory a mass of geophysical data which was not adequately familiar to geologists at that time. As a result, the elaboration of a geophysically-based plate model demonstrated that traditional geological theory was inadequate. However, on the other hand, there were too few exchanges of ideas between traditional geological views and the new plate concept. Some well-founded geological inferences were ignored, and a number of the main principles of this new theory contradicted distinct geological observations. The very existence and the possibility of movement of plates was more than once justifiably criticized, among others, by the author (Chudinov, 1976, 1981, 1985). Unfortunately, the adherents of plate tectonics, as a rule, were not able to give answers to their critics, including the present author.

The author believes that many shortcomings of plate tectonics stem from the fact that its geophysicist architects perceive geological information in too abstract a way. One of the consequences that arose from the geophysicists' geological works was that the subduction idea, which is specifically contradicted in geological facts, had been incorporated into a new concept.

The greatest disturbance of traditional geological views came from the concept of ocean-floor spreading. By now, this has developed into a well-balanced theory which is in good agreement with the results of geological and geophysical observations. It must be emphasized that the theory of ocean-floor spreading which has been developed since the early 1960s (e.g. Dietz, 1961; Hess, 1962; Vine and Matthews, 1963) had in the main been worked out by the time of the emergence of the new global tectonics as a hypothesis on simultaneous movement on the Earth's surface of several large continental-oceanic lithospheric plates. So the ocean-floor spreading theory was not a brainchild of the plate concept, as is sometimes asserted; on the contrary, the plate hypothesis itself came into being as an attempt at coordinating the already available spreading theory with the idea of the constant Earth's size (Morgan, 1968; Le Pichon, 1968). This was achieved by supplementing the spreading model with a purely speculative subduction idea, which subsequently became the weakest part of the whole plate tectonics concept.

The idea of oceanic spreading, obviously, suggests a quite different origin for the oceanic crust, and the upper mantle that underlies it, than for continental ones. We might expect that plate tectonics — the global theory with the incorporation of a spreading model — would give, in parallel with the oceanic lithosphere, an explanation for the origin of continental lithosphere. Such an explanation, however, is absent in plate tectonics and is not considered even in the works of those of its supporters who are devoted to a study of the origin of mountains (e.g. Dewey and Bird, 1970a, b). This may be a consequence of insufficient familiarity with the full range of principal geological problems, in particular, those concerning historical geology. The emergence of the continents in the course of the Earth's history, as well as their partial destruction, is proved from many pieces of geological evidence, so that hypotheses on the possible heterogeneity of the initial Earth alone cannot solve the question. No explanation is given even of initial origin of the plates as unified formations comprising both the oceans and the continents obviously created in different ways.

One new principle advanced in plate tectonics was the replacement of a relatively thin Earth's crust as the most geologically significant outer envelope by a lithosphere of much larger thickness. According to more traditional geological views, the Earth's crust above the Mohorovicic discontinuity was considered to be the main receptacle of tectonic movements and structures. This opinion should be repudiated, as the distinct character of the Mohorovicic discontinuity is called into question by geophysical investigations of recent years in many regions. The complex character of the boundary between the crust and the upper mantle, as well as often gradual transitions of one to another, has been elucidated in a number of geophysical works (e.g. Zharkov, 1978; Moiseenko, 1981). Indeed, a shell of tens of kilometres thickness is too small compared with the Earth's radius and also with the sizes of the largest oceanic and continental structures (as sheets, platforms, etc.) Besides, the idea of the universal nature of the Mohorovicic discontinuity, as well as some other depth boundaries embracing all the Earth, also seems to need revision. It is unlikely that a surface of such a relatively uniform nature, when it is down beneath the oceans, the continental platforms, and the mountains of post-oceanic origin within the continents, would have a uniform mode of origin.

Thus, a return (as this question has a long history) to the notion of a lithosphere thicker than the Earth's crust can be considered as an advancement of plate tectonics. However, the thickness of the lithosphere admitted in plate tectonics corresponds to the layer above the top of the asthenosphere, and it still does not seem to be enough. Firstly, the asthenosphere is not extended everywhere — in particular, it is usually absent beneath the continental sheets; on the other hand, in some regions there are several low-velocity seismic layers (for example, at the eastern borders of Asia — R.Z. Tarakanov, 1978; also reported in other works) which are usually considered to be criteria for recognition of the asthenosphere. It is obvious that the existence of a single surface defined as the top of such a non-universal formation can hardly be possible in reality; hence, the very existence of the surface on which the plates slip remains unproved. Secondly, the close values of the lithosphere's thickness beneath the continents and oceans claimed in plate tectonics (110—130 and 75 km respectively — Le Pichon *et al.*, 1973, and others) seem to contradict seismotomographic data which testify to the distribution of inhomogeneities in the lithosphere, including large subhorizontal ones, beneath the continents down to depths of several hundreds of kilometres, while beneath the oceans down to several tens of kilometres only.

The most plausible and useful suggestion seems to be that of H. Benioff (1954) according to which the upper shell of the Earth (which he designated as the crust) should be considered down to the base of the upper mantle: he based this on a criterion of its preserving its elastic properties as manifested in the depths of distribution of earthquake foci within seismofocal zones (which later were called after him). As the rigidity is a product of elastic modulus multiplied by sectional area, this is also the depth of preservation of rigidity properties in the lithosphere, or, more accurately, in its continental part (as Benioff considered also). The recent seismological, especially seismotomographic data (Ryaboi, 1979, Moiseenko, 1981, Sollogub, 1986, Spakman *et al.*, 1989, Delonge *et al.*, 1994, and other publications) fully confirm this conclusion while showing the layer's inhomogeneities, i e. the structure fragmentation, beneath the continents down to several hundreds of kilometres. The deepest fragmentation, and therefore the lithospheric consolidation, is probably connected with the ancient platforms within the continents.

Thus, the thickness of the lithosphere beneath the continents should be counted up to 700 km (the maximum known depth of earthquake foci is 730 km); this is the sum of the Earth's crust and the upper mantle together. An understanding of the lithosphere as consisting of the whole upper mantle together with the crust was earlier advocated by some tectonists, for instance, by M. A. Usov (1950); as has been said previously, today such a point of view is finding its verification. In this book, this is the understanding of the term `lithosphere', including the corresponding notions on its thickness beneath the continents. Undoubtedly, the structure of such continental lithosphere is rather complex and includes large inhomogeneities (among them probably more than one layer of asthenosphere type). The oceanic lithosphere, which can equally well be designated as the crust, can be considered of the same thickness as it has been suggested in plate tectonics (on the same basis of the earthquake foci distribution), i.e. of 70—100 km.

To conclude, a few words about the terminology used in this book. The concept of plate tectonics, which has become predominant over the past few decades, has developed its own new terminological system which makes use of the traditional geological terminology to only a limited extent. Without doubt, this has resulted in the loss of some ideas of traditional geology related to its major achievements. One of them is the idea of counterpoising of those parts of the Earth's lithosphere that are most resistant to deformations (the platforms or cratons) against the geosynclines (or mobile areas). While using the notions of plates covering entire continents and parts of oceans, rather than the continents or oceans themselves, plate tectonics overlooks this contrast; indeed, the logic of unified rigid plates contradicts the possibility of parts inside them that differ sharply in their mechanical properties. Some attempts were made to find a place for geosynclines in the scheme of plate movements, but the concept of platforms or cratons was ignored. One can argue whether the term `geosyncline' is adequate and whether it would be worthwhile instead to substitute the terms `paleoseas' and `paleooceans', with subdivision within them into genetic types. However, its advantage is that it allows delimitation between present-day and ancient objects studied by geology, although such delimitation may seem to be too sharp. Obviously, the term can be of further use; in this book, `geosynclines' imply those parts of the Earth's surface with a foundation of mainly mafic composition which is rather susceptible to tectonic deformations (but not necessarily from interior causes only). These structures are usually manifested in the relief as deep seas filled with sediments of different thickness depending on the surrounding conditions. Geosynclines are contrasted with ancient platforms (in the text, also `platforms') which are stable, most resistant to tectonic deformations, and are parts of the lithosphere, normally constituents of dry land. The platforms feature ancient (Archean and Old Proterozoic) foundation which is sharply different from the overlying subhorizontal (Late Proterozoic and Phanerozoic) sedimentary cover in the degree of its metamorphism and the intensity of deformations. The platform composition includes both areas where the ancient foundation is overlain with this cover, and other parts where such cover is absent, i.e. shields. Judging from the behaviour of ancient platforms throughout geological history, they are much more rigid megastructures than are geosynclinal and oceanic ones. In this work, the author prefers the term `platform' to the term `craton' according to the tradition of the Russian literature. The term became generally used after long investigations made within the Russian, or East-European, and Siberian platforms. It is closely connected with other terms like `sedimentary platform cover', `platform foundation', and `platform structures', which are derived from it but which have no commonly accepted analogues derived from `craton'. Furthermore, some authors believe that `cratons' are a synonym for shields, which does not correspond to the sense of `platform'. The latter term is also more objective as it contain no hypothetical elements (while `craton' implies source of force). The author realizes that this use of the term `platform' may cause some slight confusion to readers, as a number of western authors call platforms some moderate size localities on the Earth's surface (in tens instead of thousands of kilometres) which are characterized by subhorizontal occurrence of sedimentary beds. But such a comprehension of the term can be easily substituted for the other.

SEISMOFOCAL ZONES AND METHODS FOR THEIR STUDY

Active boundaries of the continents and oceans, together with the near-oceanic mountain ranges, island arcs, the volcanic 'fire ring', and the belts of most intense earthquakes have long attracted the attention of geologists. However, the most important feature of their deep structure, which is the occurrence beneath them of the deep zones of earthquake foci sloping downwards from the oceans towards the continents and going down to a depth of hundreds of kilometres, has become gradually apparent as recently as the 1930s. In the 1920s, it was found for the first time by K. Wadati (1928-35) that there exist deep-focal earthquakes with foci at depths down to 700 km. This study showed that they are associated mainly, and deeper than 200—300 km exclusively, with the framing of the Pacific and the Sunda (Java) arc in the north-east of the Indian Ocean. In the most detailed research of the area of the Japan island arc, completed by the early 1930s, Wadati revealed a successive increase in the greatest depth of earthquakes while moving from the continent to the ocean . It was S. Visser (1936) who first suggested the existence of single sloping zones submerging from the oceans to the continents as locations for the foci of deep focal earthquakes. This treatment of seismological evidence initially met with a critical response, in particular by such authoritative seismologists as B. Gutenberg and Ch. Richter. Further researches dispelled doubts as to the existence of single sloping zones containing deep focal earthquakes, although the shape of these zones proved more complex than a flat layer corresponding to a certain surface.

In the former Soviet Union, information on marginal oceanic seismofocal zones became widely known after the publication of the papers by A. N. Zavaritsky (1946) and N. S. Shatsky (1946). The paper read by Zavaritsky to the first post-war session of the Department of Geological and Geographical Sciences at the USSR Academy of Sciences was devoted to a review of literature on the existence of seismofocal zones and was called 'Some facts to be taken into account in tectonic constructions'; the paper by N. S. Shatsky dealt with the issues of the origin of seismofocal zones and their incompatibility with A. Wegener's concept of continental drift.

From the 1950s, attention was drawn to seismofocal zones by a series of papers by H. Benioff (1954, 1959, 1962, and other publications) which discussed the occurrence, morphology, and origin of seismofocal zones and the direction of movements along them. As for the latter, Benioff expressed different opinions (see below).

By now, extensive material has been accumulated elucidating the structure of seismofocal zones. The basic information on such zones was obtained by means of seismology, to which their very discovery was due. This information includes primarily the location and number of earthquakes. Strong earthquakes taking place in seismofocal zones are detectable by all the world's seismic stations. To record the weak, most numerous earthquakes of certain areas, a special network of stations is required; the frequency with which such earthquakes occur allows a great number of determinations to be obtained almost continuously in relatively brief periods. The position of earthquake foci allows determination of the occurrence, sizes and shape of seismofocal zones.

Wide use is made of methods for detecting space orientation of strains causing earthquakes and of directions of displacements along the rupture planes supposed to be at the earthquake's focus. The results of these determinations are more ambiguous than the data on the disposition of the foci of the earthquakes themselves. Some grounds for these determinations will be considered below. The notion of the nature and direction of strains acting at the earthquake focus depends on the selection of a model of the mechanism causing an earthquake, and several such models have been proposed. Depending on what model is selected, the accepted direction of the vectors of the acting strains may vary by 45 degrees. Conclusions as to the direction of tectonic forces may be still more ambiguous. Thus, if a common model is selected which proceeds from the existence of a three-axis ellipsoid of strains, in most cases, one cannot determine unequivocally whether a displacement was caused by the action of compression or extension forces. If one does not use a special technique, which in fact is not always used, then this geologically important conclusion can be of dubious reliability: according to the theory, the displacements along two oblique sections of the strain ellipsoid are equally probable, and the conclusion as to along which of these planes the actual displacement took place is drawn without any strict grounds and proceeds from indirect evidence. These circumstances are often ignored in the geological interpretation of geophysical evidence.

Seismological investigations proper can be supplemented with direct observations over earthquake displacements found on the Earth's surface. However,such observations can be carried out only infrequently, but their generalization is of obvious interest. For seismofocal zones as a whole, such a generalization does not seem possible; but there are works available in some of the earthquake regions or certain groups of them (for example, the works by L. M. Balakina pay considerable attention to tsunami-making earthquakes, e.g. Balakina, 1972).

A number of data on structure and physical properties of seismofocal zones and adjacent volumes of the crust and mantle can be derived by determining the velocities of waves passing through them as well as from the capacity of different sections of the crust and mantle to conduct earthquake waves without absorbing them (called earthquake quality and designated `Q'). These methods have yielded evidence of considerable heterogeneity of the matter within the volumes enclosed by seismofocal zones, in particular, on distribution of masses of low quality in both wings, especially in the hanging one, and on the presence of a bed-shaped zone of a higher quality which is close to the main ocean-facing boundary of the seismofocal zones. A more detailed study reveals considerable heterogeneities inside those formations too.

Seismological methods have yielded basic evidence on both the existence and the shapes and sizes of seismofocal zones. However, the issues of the locations and directions of tectonic movements in seismofocal zones are not resolved by these methods.

Quite important evidence on the nature of seismofocal zones has been obtained from gravimetric investigations. They have established that in the regions where seismofocal zones outcrop onto the surface there exist the world's largest linear gravitational anomalies which testify to disturbances of isostatic balance in connection with existence of seismofocal zones and, obviously, indicate continuous tectonic movements which cause such disturbances. But the gravimetric data cannot determine the direction of such movements.

Magnetometric techniques used in the investigation of seismofocal zones have so far been of less importance compared with seismological or gravimetric techniques. The external side of recent deep-sea trenches sometimes have stripe-type magnetic anomalies — but sometimes they have none. They are normally of less stable character compared with stripe-type anomalies within mid-ocean ridges. With respect to the strike of trenches and seismofocal zones, they are either oblique or subparallel. They are often interpreted as being spreading anomalies which migrated towards the trenches and are thought not to be connected with seismofocal zones in their origin, though this is not certain.

Geologically, evidence on seismofocal zones cannot be considered complete enough unless their nature as tectonic structures is understood, i.e. their location, sizes and shapes of the surfaces of tectonic displacements, properties of structures divided by them, and the direction and amplitude of movements. The issue of direction of movements is especially important since it determines a proper understanding of the processes which develop in connection with seismofocal zones. The possibility of solving this issue by geophysical techniques seems to be limited. However, such an answer can be suggested from geological observations. All the circumstances testify to the fact that the tectonic displacements connected with seismofocal zones are large and, consequently, should leave clear traces in specific properties of geological structures and formations which accompany such seismofocal zones. These can be studied in two ways. One consists in studying ancient geosynclines (as analogues of the recent marginal seas) whose development was connected with paleoseismofocal zones (to cite one example, one can mention the Koryak geosyncline in the Russian North-East whose geological history is considered in Chapter IV). The other way is to use materials from deep-sea drilling of the modern ocean basement, where recent seismofocal zones outcrop onto the surface.

Occurrence. Deep seismofocal zones are widely, but not quite evenly, distributed over the Earth's surface (Fig. 1). Normally, their outcrops are accompanied by deep-sea marginal oceanic trenches which, thus, indicate the existence of seismofocal zones. Seismofocal zones are most developed on the fringes of the Pacific, covering about two-thirds of its perimeter. Starting from the Kermadec—Tonga Island Arc in the south-western Pacific, seismofocal zones extend in a continuous chain along its western and north-western fringes as far as the point where the Aleutian Island Arc joins the North American continent, stretching over almost half the circumference of the ocean. Here, the outcrops of seismofocal zones onto the surface are marked as a chain of deep-sea trenches behind which, towards the continent, one can find the chains of island arcs — festoon-like `garlands' of the western and north-western Pacific — and still farther behind them there are vast marginal seas like the Fiji, Coral, Philippine, and Bering Seas, and others. Seismofocal zones are normally inclined towards the surrounding continents and, thus, island arcs are situated over them. The Philippine Sea is bordered by seismofocal zones and accompanying deep-sea trenches on both sides: the first seismofocal zone (external) comes out onto the surface in the area of the Mariana Trench and Mariana Island Arc; the second is along the western sea border marked by the Philippine Trench and Philippine and Ryukyu (Nansei) island arcs. Both seismofocal zones are inclined to the same side — to the west; thus, the deep structure of the Philippine Sea inside two subparallel inclined seismofocal zones, as well as other depth structures of this part of the Pacific, are determined by a pair of seismofocal zones inclined to one side, and they are clearly asymmetric.

There are also opposite dip directions of pairs of closely disposed seismofocal zones. These refer to the area between New Guinea and the Tonga arc whose length is about equal to that of the Philippine Sea system of two seismofocal zones inclined to the same side. Between the islands of New Guinea and Tonga, there are two subparallel chains of trenches and accompanying seismofocal zones, although interrupted. These chains are remarkable in that their subparallel seismofocal zones dip towards one another. The seismofocal zones in the northern chain dip from the ocean to the continent (in the south-western direction), in common with most other zones, but more continuous seismofocal zones of the southern chain are inclined from the Australian continent towards the Pacific (to north-west), i.e. in the direction opposite to the usual one. Between these two seismofocal zones, inclined to meet each other, there are the islands of the Bismarck Archipelago (New Britain, and other islands), Solomon Islands and New Hebrides. The counter dipping of seismofocal zones (i.e. the counter underthrust of the plates) in a region of New Hebrides with their outcrops onto the surface approaching each other as close as 500—800 km, especially the backward dip of the southern chain of seismofocal zones, presents great difficulties to the adherents of subduction. To apply the subduction concept in this case they have to assume perfectly anomalous conditions for plate collision.

The described island arc part of the Pacific margins can be referred to as the Kermadec—Aleutian sector. As seen from the above, this sector is characterized by development of island arcs, marginal seas and the existence of complicated doubled systems of seismofocal zones with similar or counter dips. The depth penetration of seismofocal zones and their inclination, as determined by the distribution of earthquakes, prove to vary from place to place. Depths of 600—700 km are reached by the seismofocal zones in the rear of the Tonga arc (to the south of the Kermadec arc, the zone depth decreases); beneath the Philippine and Mariana arcs they are 600 km deep, in the rear of the Japan, Kuril, and Izu—Bonin arcs they are 500 km deep;

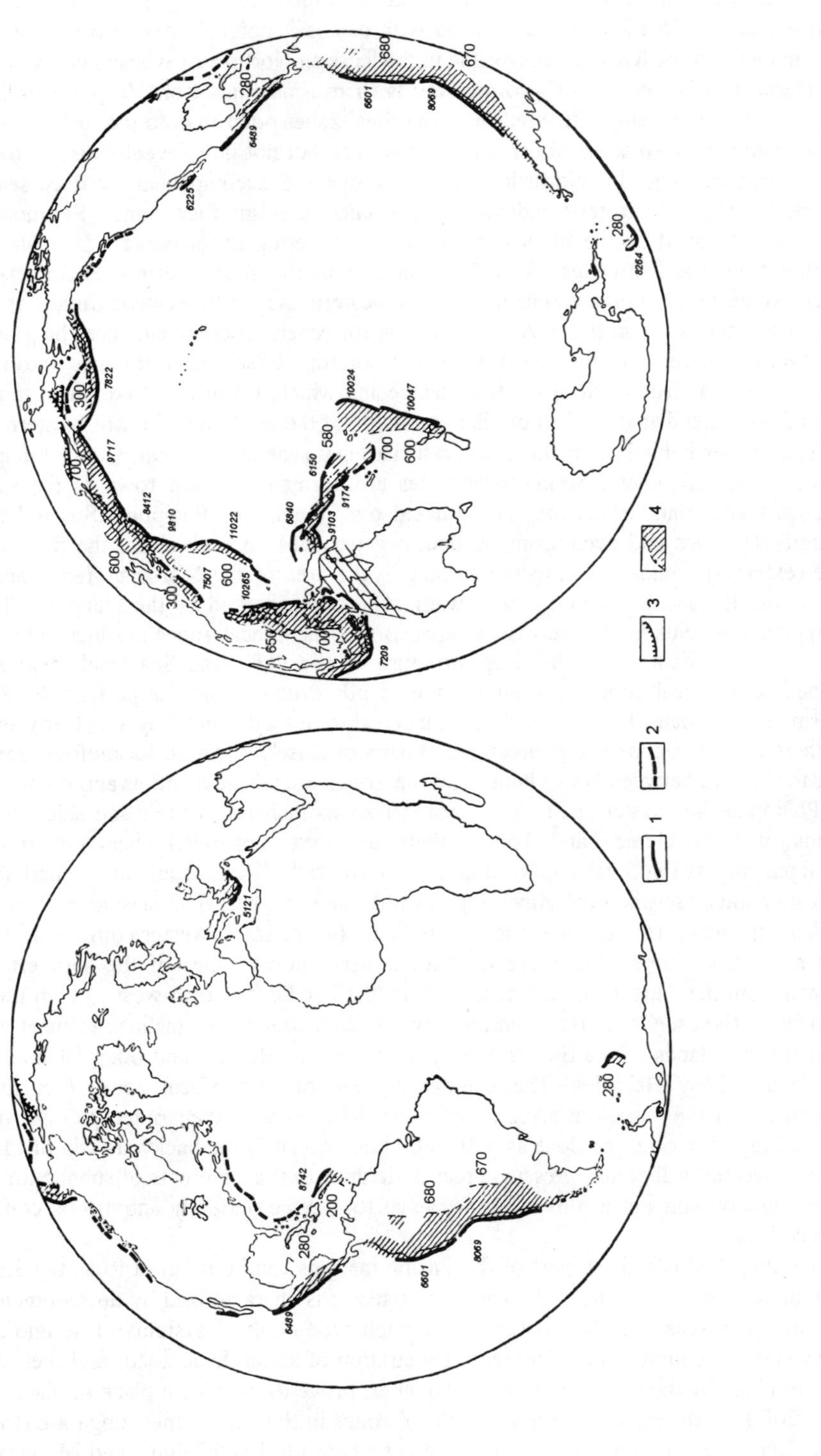

Figure 1. Seismofocal zones. 1—deep oceanic marginal trenches manifested in present relief; 2—trenches buried by Meso-Cenozoic sediments; 3—location of some paleo-trenches in Mezo-Cenozoic folding provinces in the Pacific surroundings; 4—areas of projections onto Earth's surface of recent seismofocal zones; numbers: three-figure—maximum depths of seismofocal zones, (km); four- and five-figure—maximum depths of trenches, (m).

the same depth can occasionally be found beneath the Solomon and New Hebrides Islands (but mostly they are shallower); beneath the Aleutian arc the depth of the seismofocal zone is not greater than 250—300 km. The average inclination of the seismofocal zone beneath the Mariana arc is 70° and in its lower part it is nearly vertical according to the currently available data; dips of up to 70—50° are typical of a considerable part of the seismofocal zones beneath the Philippines and New Hebrides. Beneath the Kuril—Kamchatka arc and the Sea of Okhotsk, the seismofocal zone dips by an order of 50°; beneath the Aleutian arc and the greater part of the Japan arc by 45°; and at some sections beneath the Japan Islands and Sea of Japan by up to 35°.

To the east of the Kermadec—Aleutian sector, the Pacific fringing is represented by the coast of North America where modern seismofocal zones discontinue. However, during the relatively near geological past they existed there, which is testified to by the trenches buried by sediments but detectable in geological and seismological investigations.

A new chain of deep-sea trenches and seismofocal zones which dip towards the American continents starts in the form of a small trench off the western coast of the California Peninsula to stretch along the coast of Central America and, following its disruption near the Isthmus of Panama, further along the greater part of the South America coast. This second sector of the Pacific fringing characterized by development of seismofocal zones can be referred to as the California—South American sector. In its extension, it is about 1.5 times the length of the Kermadec—Aleutian sector. At its ends, in the area of California and near Tierra del Fuego, seismofocal zones are not clearly distinct, but they are well pronounced in the central American part and along the rest of the South American coast. The seismofocal zones of the American coast are of a simpler structure than those on the opposite side of the ocean: they are everywhere single, i.e. they join each other only along the strike and dip only to beneath the continent. The depth of the seismofocal zone beneath Central America reaches 250—300 km with a dip of about 45°. Beneath the South American coast, the single seismofocal zone stretching from South Ecuador to the Chonos Archipelago in the south of Chile penetrates down to a depth of 600-650 km; it is about 5000 km long which makes it the Earth's longest seismofocal zone. Its dip is occasionally more gently sloping in the upper part and steeper in the lower. It is this fact that gave grounds for H. Benioff's (1962) opinion on the regularity shared by all seismofocal zones which consists in their more gentle dip in the upper part with a steeper dip in the lower regions.

According to Benioff, a change of the dip angle takes place near the lower boundary of distribution of intermediate earthquakes at the depth of 300 km; in South America the dip angle of the seismofocal zone reaches 23° in the upper part and 50-60° in the lower (Benioff, 1962). More recent data (e.g. Krasny and Fedynsky, 1976; Sacks, 1977) evidences that a change in the zone's steepness does not occur everywhere and a change of the dip angle, where detectable, is often at the shallower depths.

The largest seismofocal zone beyond the surroundings of the Pacific is the zone beneath the Sunda arc starting off from the Sunda Trench and submerging to beneath the Indonesian Archipelago. The depth of this seismofocal zone reaches 700—720 km with its general dip about 45°. The seismofocal zone of the Sunda Trench dips to meet, at depth, the seismofocal zone of the Philippine Trench. Their outcrops onto the surface, shown by trenches, come closer to each other in the south-east direction to a distance of 2000 km and the projections of lines of their grounds with deepest earthquakes draw together in the central part of the structure (of 3000 km width) to a distance of less than 2000 km and on the south-east up to 1000 km. So the Sunda and Philippine seismofocal zones form, in essence, a system of two zones which dip to meet each other, which is similar to what occurs on both sides of New Britain, Solomon Islands and New Hebrides; thus, from its disposition, one system may be considered a continuation of the other. It may be assumed that, in this case, the central part of an originally single megastructure bounded by seismofocal zones with counter dipping was deformed and destroyed during the northward motion of the Australian continent (which resulted also in the origin of the folding structures of New Guinea). Such movement is suggested in all the reconstructions that consider displacements of the continents. If the supposition is true, then the Tonga—Indonesian deep megastructure, stretching from the Tonga basin and New Hebrides to the south-east of Asia, still existed at the beginning of the Cenozoic. Deformation of such a megastructure can account for the shape and origin of a number of patched curved trenches and their accompanying seismofocal zones near New Guinea (in the Banda Sea and off New Britain Island). But independently of this assumed single structure, the counter dip of the Sunda and Philippine zones challenge subductional views in the same way as does the one in the area of New Hebrides.

A kind of replica of the conditions in the Indonesian region is the distribution of seismofocal zones in the area of the Caribbean basin. West of it, behind a relatively narrow Central American isthmus, is the above-mentioned Central American seismofocal zone; in the east, adjacent to the northern part of the Antilles arc is the Puerto Rico Trench with its accompanying south-inclined seismofocal zone, the latter going down to a depth of 200—300 km to continue southeastward replicating the curve of the Antilles arc. The Puerto Rico—Antilles seismofocal zone appears, similarly to the Sunda seismofocal zone, on the nearest border of the ocean neighbouring the Pacific (in this case, of the Atlantic ocean), where the two oceans are closest to each other. They are divided by a dam, probably of disintegrated fragments of the continental lithosphere, which repeatedly alternate with the ocean-type basins. In both cases, the seismofocal zones which are on the border of the Pacific and the adjacent oceans are of a counter dip.

The Puerto Rico—Antilles seismofocal zone may be a remnant of an older and extended zone which outcropped onto the surface along the Atlantic coast of North America and the north-eastern slope of the Bahamas. Its existence may be inferred from the remnants of the buried trenches which stretched from the foot of the Bahamas to the Great Newfoundland Shoal —trenches which are detectable by seismic investigations (e.g. Emery, 1965; Sheridan, 1974).

A seismofocal zone with earthquake depths down to 300 km accompanies the South Sandwich Islands (South Antilles Island Arc). This island arc, which is clearly convex towards the east and morphologically similar to the Antilles arc of the Caribbean Basin, is accompanied on its east side by a conformly curved trench near which starts the above-mentioned seismofocal zone that dips eastward towards the Scotia Sea. This is the second of the existing, rather short (not longer than 1000 km), recent seismofocal zones associated with the eastern margin of the Atlantic Ocean.

Finally, let us consider the seismofocal zones on the fringes of the Mediterranean. Most uniform and pronounced among them is the zone accompanying the Aegean island arc and the Hellenic Trench which is at its front. The seismofocal zone slopes northward here to beneath the island arc, penetrating to a depth of 250 km and sloping at 45°. Its frontal length is about 800 km. A fragment of steeply dipping seismofocal zone may also exist to the south of Calabria and Sicily where, near the Strait of Messina, some earthquake foci are as deep as 400 km. Deep earthquakes of depths down to 500-600 km are also known beneath southern Spain.

As may be seen from the above description, one can identify two main types among recent seismofocal zones with respect to their correlation with the largest elements of the earth's relief: those beneath the island arcs (which are most common) and those immediately beneath the continental margins (within the California—South America sector of the Pacific fringe). The island arcs are normally accompanied by seas of a marginal type which are partially or wholly situated above the seismofocal zones (the South Okhotsk Basin in the Okhotsk Sea and the Sea of Japan are entirely above the seismofocal zones). The steepest dips of the seismofocal zones are found beneath the narrow island arcs with a low-standing basement whose surface is mainly submerged under sea level (Mariana, Tonga). The depth of the seismofocal zones varies greatly and seems, to some extent, to be influenced by the sizes of the island arcs, mainly, their extension: the longer the arcs, the greater are the depths, and the comparatively shorter arcs produce shallower depths. In the California—South American sector, the depth of penetration of the seismofocal zones may be associated with the width of the adjacent part of the continent. The greatest depth of the earthquake is found in the central part of the South American continent, within North Argentina—Bolivia—West Brazil. The gentlest (on the whole) slope of the seismofocal zone was also found here. The shallowest depth of the seismofocal zone is beneath Central America, where the continent width is rather narrow.

A characteristic morphological feature of the island arcs accompanying seismofocal zones is their oceanward convexity. The curvature of this convexity proves to be quite variable. Using the ratio of the arc height to its span on the Earth's surface, this parameter varies from 1:10 on the Kuril chain to 1:3 or 1:2 or more in the Mariana arc and both Antilles arcs. It was established that the observable convexity of the most strongly curved arcs geometrically cannot be accounted for by intersection of the plane seismofocal zones with the Earth's surface as suggested by some researchers; so the convexity of island arcs as well as of their accompanying trenches should depend on the corresponding convex shape of the structures that accompany seismofocal zones. An exception with respect to convexity of island arcs is the Kermadec—Tonga rectilinear island chain and trench. The lines of the seismofocal zones outcropping onto the Earth's surface in the California—South America sector do not fall in any definite regularity but are, for the greater part, rectilinear, although the curve of the Andes and the entire coastline off the Peruvian Trench (opposite the Amazon basin east of the Andes) can be interpreted as a bend similar to that of the island arcs. Some curvature towards the Pacific is also characteristic of the northern part of the Central America Trench together with its corresponding coast.

Structure of upper parts of seismofocal zones. Seismofocal zones are identified on the basis of the occurrence of earthquake foci and correspond to the space occupied by those foci. This can be very great: for example, some seismofocal zones are as wide as 600 km or wider in their upper parts: so they cannot be identified with certain geological boundaries. Since seismofocal zones cover large sections of the lithosphere, they themselves may include several great deep structures with their boundaries, or, alternatively, a single great structure, or a part of it. The upper parts of seisomofocal zones lie at depths of less than 300 km, where these zones are the widest and most complicated in their inner structure.

The shape of the cross-section of the upper part of seismofocal zones can be classified into three types, with a gradual transition between them.

The first type of seismofocal zone is of a shape close to sheet-like, which is typical of many lateral sections passing through the Japan Island Arc (Fig. 2). These sections occasionally display two parallel belts of clustered earthquake foci which seemingly mark two parallel planes (Fig. 3). These data on the occurrence of earthquakes in Japan and some other places contributed to the widely accepted idea of a rather clear-cut sheet-like shape for seismofocal zones, since Japan is a classic country for their study. However, as one can see from the

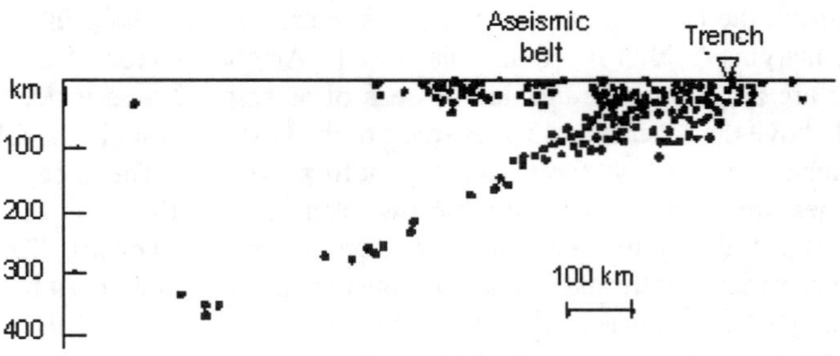

Fig. 2. Seismofocal zone beneath Japan (after T. Yoshii, 1977).

illustrations (Figs 2 and 3), in the upper lithosphere, over 50 km, the area of occurrence of earthquake foci widens considerably, especially in its near-surface portion, i.e. in the earth's crust proper, up to 400-600 km. Hence—and this is an important point—the expansion tends mainly to the landward side and only insignificantly towards the ocean (taking the trench to be a boundary between land area and the ocean).

The second type of seismofocal zone can be found near Kamchatka and some other arcs (for example, near the Mariana arc — Krasny and Fedynski, 1976). The Kuril—Kamchatka arc is a continuation of the Japan arc, where one can see sections of the seismofocal zone similar to the Japan type described above, and those of an intermediate third type. In many cross-sections through Kamchatka, the seismofocal zone differs from a sheet-like shape and looks like an upturned bell widening at the top (Fig. 4). The width of this funnel-shape form varies from tens of kilometres at a depth of 200—300 km gradually growing to reach 400—500 km near the surface. The volume of distribution of the earthquake foci is disposed mainly on the continental side of the trench and barely includes the crust and lithosphere on the oceanic side of the trench. Beneath the trench and its outer oceanic-side slope, the earthquake foci are distributed down to a depth of 100 km and not deeper, so the level at which they appear is higher than that where the seismofocal zones start widening on the continental side of the trenches. One should note that the frequency of earthquake foci differs from one side of the Kamchatka `funnel' to the other: it is much greater in the part facing the ocean (mainly beneath the continental slope). As seen from Fig. 4, the funnel border on the oceanic side is much more clear-cut than it is on the side facing the continent; the latter is blurred due to a gradual decrease in the number of earthquakes towards the continent.

The third type of seismofocal zone is characteristic of South America and its cross-section is of a triangle shape widening and blurring towards the continent (Fig. 5). The boundary of the volume of distribution of earthquake foci facing the ocean, similar to other cases considered, is clear-cut and accompanied by the greatest number of foci. The number of earthquakes gradually decreases beneath the continent. Most earthquake foci fall at depths shallower than 200—250 km. The earthquakes occurring within this depth interval spread inside the continent to a distance of 800 km. The sections shown in Fig. 5, similar to those in Fig. 2, omit the earthquake foci deeper than 300 km, i.e. those which are deep-focus earthquakes proper (they are less numerous and reach depths close to 700 km; the epicentres of the deepest earthquakes beneath South America are 750—1000 km off the coast and the trench).

All three types of seismofocal zones considered here share a clear-cut boundary of the volume of distribution of earthquake foci from the ocean side, and their greatest concentration is near this boundary. The variation in the section shape of seismofocal zones, depending on

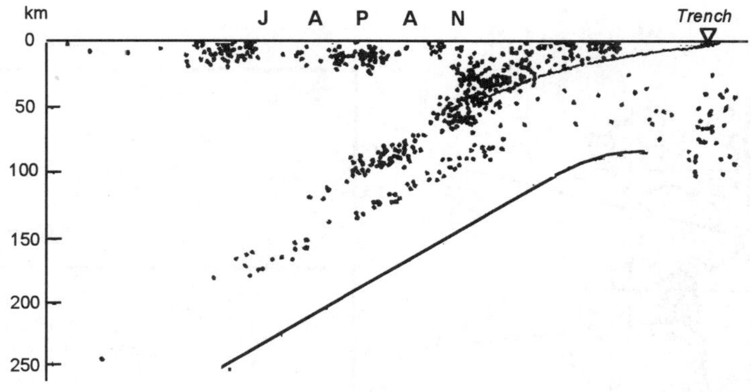

Fig. 3. Double-planed seismofocal zone beneath Japan (after A. Hasegawa *et al.*, 1978).

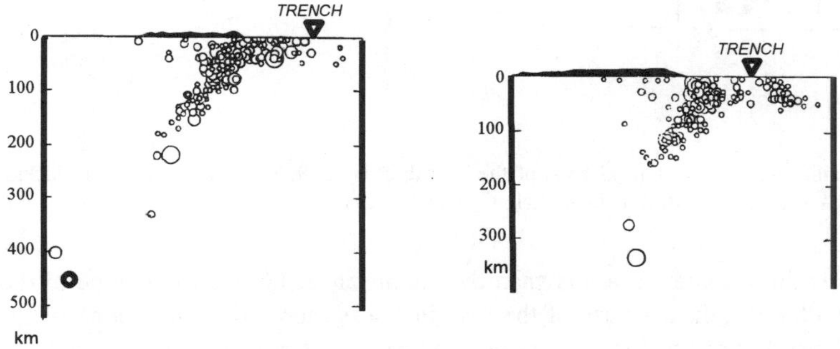

Fig. 4. Two sections of seismofocal zone beneath Kamchatka, (after R.Z. Tarakanov, 1981). Horizontal and vertical scales are equal.

the kind of distribution of earthquake foci towards the continent, may be attributed to the basic features of the structure of the lithosphere adjacent to the ocean. It may be suggested that a relatively narrow width of seismofocal zones beneath the Japan Island Arc (as well as beneath other island arcs) is due to the marginal seas of an oceanic-type base which are behind the zones. An analysis of the location of the earthquake epicentres behind the Japan and Kuril-Kamchatka arcs (Fig. 6), which is one of the best studied seismic areas of the Earth's surface, shows that the southern half of the South-Okhotsk deep-sea basin and the deepest parts in the Sea of Japan adjacent to the Yamato Rise on the north and south display `gaps' within which the earthquake epicentres are either absent or their number drops abruptly. Obviously, some volumes of the lithosphere beneath these seas are aseismic ones, and such `gaps' can also be seen in other marginal seas. On the other hand, the width of the band where relatively shallow to intermediate earthquakes occur depends on the width of the island arc. This band is narrowest over the Kuril arc where it spreads covering the island arc slope of the trench and the arc proper which reaches 200—250 km. In the Central Kamchatka, the width of the band of shallow earthquakes increases up to 450 km and wider, and that on the Japan islands up to 400—600 km. To the east and towards the continent, shallow and intermediate focus

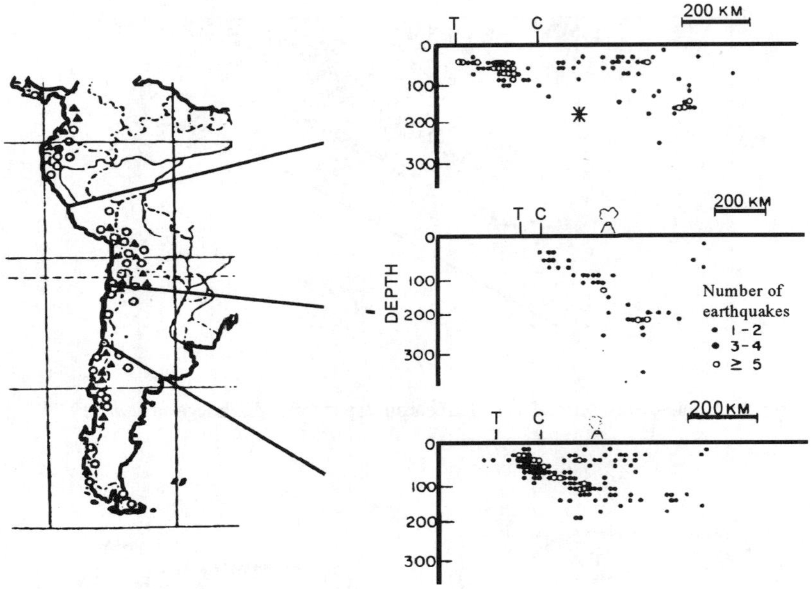

Fig. 5. Sections across seismofocal zone of the Pacific coast of South America (after I. Sacks, 1977). 1-2—volcanoes (1—active, 2—extinct); T—trench, C—Pacific coast.

earthquakes do not stretch farther than the continent-facing (internal) slope of the island arcs whereas within significant parts of the marginal seas there are almost none of them. Judging from the distribution of the epicentres (and hypocentres) one can conclude that: (1) the lithosphere beneath the marginal seas is much less seismically active than it is at the base of the island arcs, (2) most earthquakes develop at the base of the island arcs including their mantle part.

The features noted above of the distribution of earthquakes indicates an alternation of lithospheric masses with different elastic properties. If one considers the generation of marginal seas, as a whole, as a consequence of destruction of the continental lithosphere by stretching and disrupting (with many arguments in favour of this in recent investigations), then the cause of such an alternating pattern is attributable to separation of the most elastic fragments of the continental lithosphere and filling of the space between them by low elastic masses which are supplied from depth and which do not favour generation of a large number of earthquakes. The first type of the above-described seismofocal zones observed in the region of the Japan arc can be viewed as a type of shortened seismofocal zone whose development is not full enough due to heterogeneous lithospheric structures and partial distribution of relatively inelastic masses. These include the basement of marginal seas and the roots of volcanic chains whose poor seismicity has been mentioned in a number of geophysical researches (and make up a considerable part of the foundation of the Japan Islands). One can suppose that beneath the Japan Islands only the most pronounced constituent parts of the seismofocal zone, in the form of earthquake foci clusters near its ocean-facing boundary, are mainly displayed.

A far propagation towards the continents of the earthquake foci in other types of seismofocal zones, with their gradual disappearance, can be explained by comparatively high elasticity and consolidation of the sufficiently continental lithosphere being enclosed by the seismofocal zone. The intermediate second type of seismofocal zone observed in Kamchatka displays this feature more vividly than in Japan which is compatible with the continental-type lithosphere more widely spread in this area and making up not only the foundation of the Kamchatka peninsula but also a part of the adjacent foundation in the Sea of Okhotsk. But here, too, the lithosphere is heterogeneous in its composition. The least disturbed continental lithosphere, of essentially platform nature, is that enclosed by the seismofocal zone beneath the South American continent. This third type seems to represent the most fully developed seismofocal zone.

Support for the opinion that the foci of earthquakes creating seismofocal zones prevail in the continental-type lithosphere is offered by the results obtained from researches of elastic properties of crust-mantle substance in the areas where seismofocal zones are developed. Figure 7 shows the sections illustrating the results of determination of seismic quality of the lithosphere beneath the Japan Islands and Pacific coast of South America. Since this seismic quality reflects the capacity of the material to partially absorb seismic waves during their propagation, its growth corresponds to an increase in the elastic properties of the material. The sections reveal that in South America, the lithosphere capacities on the oceanic and continental sides of the deep-water trench differ greatly. Beneath the continent there are masses of high quality (Q = 1000—3000) and, beneath the ocean, those of lower quality (Q = 300—500). The border between those two types of lithosphere is nearly rectilinear and slopes by about 45° towards the continent to outcrop onto the surface near the trench axis. Here, there is an obvious contact of two deep structures with sharply different properties of mantle. By comparing the quality of the section in Fig. 7 with that of the section of earthquake foci occurrence beneath South America in Fig. 5, one can ensure that the earthquake foci, which make a seismofocal zone in South America 800 km wide, are fully within the deep continental structure of high quality (elasticity) of the material of which it is composed.

The distribution of rock elasticity/Q-values beneath the Japan arc also corresponds to the suggestions made above. Similar to the case near South America, adjoining on the side of the ocean to the seismofocal zone, there is a mantle of poor quality of 400 (a high quality on the oceanic side in both cases is found only in the near-surface layer of thickness of the order of 70 km). Right from the trench and to beneath the Japan arc runs a zone, inclined at about 45°, in which the quality changes in different ways. Immediately beyond the boundary with the low-quality oceanic mantle there follows a sheet-like high-quality zone conforming with that boundary and partially coinciding with the above-mentioned sheet-like zone of clusters of earthquake foci beneath Japan; the high-quality zone is followed by a mantle of medium quality and deeper than 200 km there is a mantle of sharply decreasing quality. Higher, and nearer the surface on the continental side of the trench, high-quality crust of about 50 km thickness can be found. Then, for several hundreds of kilometres towards the continent at a depth of 200 km and deeper, there appears again a district of high-quality mantle. The figures on the section (see Fig. 7) which designate earthquake frequency also show that the earthquake foci are arranged selectively in high-quality districts of the crust and mantle.

Structure of deep parts of seismofocal zones. A gap of no earthquake foci is found in the interval from about 300 to 450 km, sometimes higher, with the top up to 200 km, in a number of the deepest seismofocal zones; but below this depth, earthquake foci again appear and make a band which is narrower than in the upper parts and seems to be of tens of kilometres, up to a hundred kilometres, and continues in the direction of the inclined ocean-facing boundary of the upper part of the seismofocal zone. At a depth of 650-700 km, in some cases, the zone of earthquake foci distribution seems to flatten out.

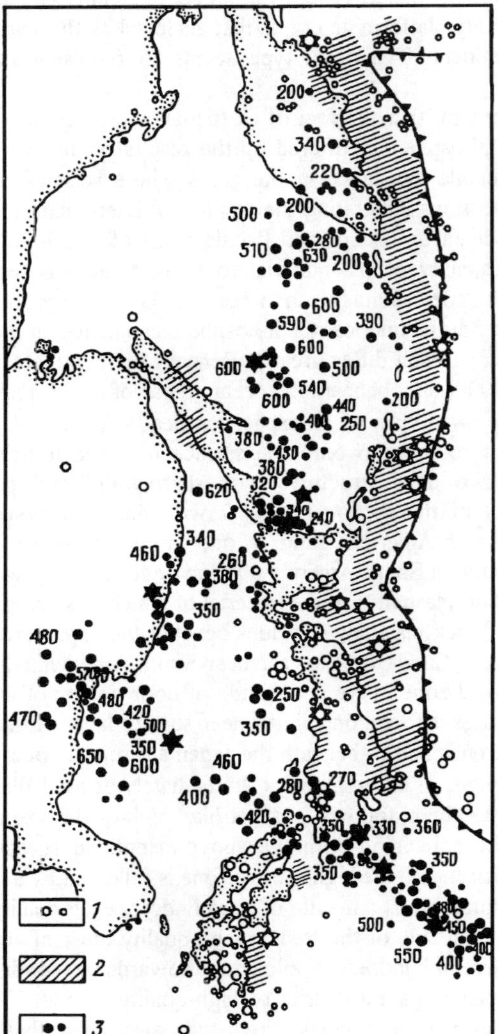

Fig. 6. Location of earthquake epicentres above Kuril-Kamchatka and Japan seismofocal zones (from 1904 to 1968), (after to R. Z. Tarakanov, 1976; simplified). 1—Earthquakes with hypocentre depths down to 200 km; 2—areas of highest density of epicentres of earthquakes with the same hypocentre depths; 3—earthquakes with hypocentre depths of more than 200 km (numbers show hypocentre depths in km). Size of circles is proportional to value of the earthquake's magnitude; sprocketed circles are earthquakes of magnitude 8 and stronger.

Some authors cast doubts as to whether the gap really exists using, in particular, data from the seismofocal zone of Japan, where it is not found everywhere. However, as has been said above, the structure of some seismofocal zones, and in particular in Japan, may show large anomalies. It is the seismofocal zone of South America, the most homogeneous, that should be taken as the standard of a seismofocal zone. For in this case, the aseismic gap is distributed over the entire zone, so its existence should be considered as universal enough.

It may be suggested that there is a thick layer in the lower part of the upper mantle, mainly beneath the ancient platforms, which is remarkable for its especially low elasticity; intersection of it by a tectonic structure causing the existence of seismofocal zones leads to the absence of earthquake foci within them along this interval. Indeed, when looking at the sections of elastic properties of the lithosphere of Japan and South America in Fig. 7, one can see that in both cases the level of the seismic gap (300-450 km) coincides with the appearance of large fields of material of sharply increased seismic quality which adjoin the main line of division of the seismofocal zone structures, mostly on its continental side; beneath South America such a field is of clearly horizontal bed-shaped form. It is possible that this layer is a more important upper-mantle structure than the asthenosphere layers distinguished higher up.

Relief of the Earth's surface. There exist two main types of macrorelief accompanying a seismofocal zone: one is characterized by development of an assembly of a deep-sea trench—island arc—marginal sea; the other type is that where the trench immediately adjoins the continent's edge, which is accompanied by a chain of mountains. The latter is referred to by some authors as the Andean type, while the former features is called an island-arc, or West-Pacific, macrorelief.

The most permanent and indispensable element of relief accompanying the seismofocal zones is that of deep-sea trenches. Their structure is similar both in front of the island arcs and along the Pacific coast of America; in general, it is uniform throughout all parts of the globe. The main difference in the trenches consists in their shape in plan (which has been mentioned already) and, to a lesser degree, in their depth. The length of trenches generally varies from 5000 km (Peruvian—Chilean Trench) to 1000 km and lesser (Puerto Rico, South Sandwich Islands, Hellenic Trenches). The longest after the Peruvian-Chilean Trench are: Aleutian Trench — 3500 thousand km; Tonga-Kermadec Trench and Sunda Trench — 3000 km; the Central American Trench and Mariana Trench (together with its northern extension Volcano Trench) — 2500 km; Kuril-Kamchatka Trench — 2000 km. The greatest depths vary from 9—11 km in large trenches of the western Pacific (Tonga-Kermadec — 10.9 km; Mariana — 11 km; Philippine — 10.3 km; Izu—Bonin — 9.8 km; Kuril-Kamchatka — 9.7 km) and drop to 6—8 km in the Ryukyu (Nansei) Trench north of the Philippine one, 6.5 km in the Central American Trench, and about 5 km in the Hellenic Trench. The Peruvian-Chilean Trench, which is divided by a rise opposite the submarine Naska Ridge, is in its northern (Peruvian) part rather shallow with a depth of 6.6 km, but in the southern (Chilean) part it is about 8.1 km deep. The depth of the Aleutian Trench is 7.8 km; that of the Sunda Trench, 7.2 km; the South Sandwichean Trench. 8.3 km; and the Puerto Rico Trench, 8.7 km. The evidence cited testifies to no direct connection between the extension and the depth of a trench. Common depths, thus, are 7—10 km.

The axial deepest part of the trenches looks either like a narrow slit if there are no significant sediments there (which is a rather frequent case—Figs 9, 13) or, with sediments (Figs 8, partly 9, 16) like a plain narrow surface from several to twenty kilometres wide, horizontal or mildly heightening oceanward. The axial band divides two distinctly different slopes of the trench—a continental or island arc slope on the one side and an oceanic slope on the other. These slopes differ in their steepness and width which makes all the trenches asymmetric in cross-section (Figs 9, 13). Let us refer to the slopes of the continental edges and

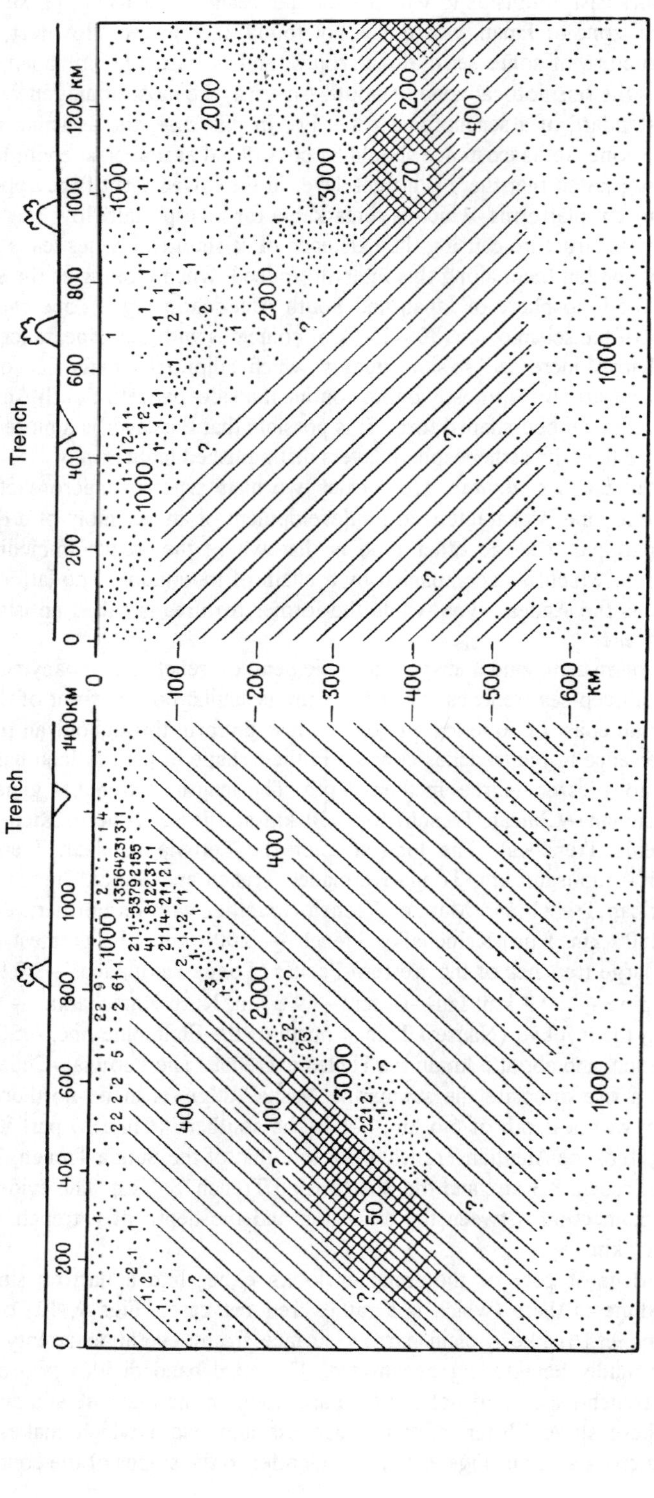
Trench
0
200
400
600
800
1000
1200 км
1400км
100
200
300
400
500
600
км
1000
2000
3000
400
100
50
70
200

Fig. 7. Seismic quality beneath Japan (left) and South America, (after Sacks, 1977). Dotted are volumes of high seismic quality (Q=1000-3000), hatched—of intermediate quality (Q=300-500), and cross-hatched—of low quality (Q=50-100). Within volumes of high quality, small digits show dislocation of earthquake hypocentres and their number. Volume over boundary of seismofocal zone (i. e. over surface of main tectonic movements) is characterized by intermediate quality beneath Japan and by high quality beneath South America. In both cases, especially beneath South America, considerable decrease of seismic quality is observed below 300—350 down to 400—450 km.

of the island arcs as continental slopes since both are similar in structure. They are much higher than the oceanic ones as they usually end up above the sea level; besides, for example in the northern Chilean Trench, their immediate land continuation is an adjacent slope of the Andes, which makes the total elevation of this continental slope over the trench floor as much as 12—14 km. Some under- and above-sea-level slopes of island arcs are also of a similar structure, but the height of their above-sea portions does not exceed 2—3 km against 5—7 km in the Andes. In many cases, including on the Andean coast, between the mountains and the submarine slope there exists a shelf of significant width which divides a submarine slope from the land mountain structures. However, even when counted from sea level, the height of the continental slope is 6—10 km and more, depending on the trench depth. The width of the submarine part of the continental slope from the trench axis to the coast line varies from trench to trench and normally reaches 75—100 km; an average angle of inclination of the surface of continental slopes is thus equal to 3—7° but occasionally it is much steeper. The oceanic slopes of the trenches are wider — usually 125—150 km, and their height is lower; therefore they are much gentler than the continental slopes and their total inclination is less than 2°. The oceanic slope of the trenches is often separated from the adjacent oceanic basin by an external oceanic swell which is a gentle rise up to 100—150 km wide and by a height above the farther oceanic floor of 500 m, less frequently of up to 1000 m. Since the usual depth of the oceanic basins reaches 5 to 5.5 km, the elevation of the axes of swells over the trench axes is within 4—5 km or less.

The continental slopes of trenches are characterized by much greater tectonic ruggedness compared with the oceanic ones. Their lower parts are the steepest — about one third of the whole slope — with an average inclination of as much as 15°, and numerous tectonic steps are accompanied by steeper and even vertical portions (Figs 10, 11, 12, 13). Higher up on the continental slopes, there appear gentle-sloped submarine terraces making one, two or even more levels. Over or between the terraces some comparatively steep portions are observed too. In different parts of the continental slopes, especially often near the foot and also often on the terraces, one can find complicating graben troughs parallel to the trend of the slopes; on the edges and in the central part of the terraces there are also some horst projections. Sometimes the horst crests may form a kind of local cordillera on the external side of the terraces.

The origin of submarine terraces has generated interest from many researchers who express an opinion on their either tectonic or landslide origin. It seems certain that landslides do exist, and sometimes very large ones, considering the existing steepness of the continental slopes and constantly repeated earthquakes. However, the largest forms of relief of continental slopes seem to be created mainly by tectonic processes, which is certain from the images of bedrock structures in numerous seismoacoustic profiles, and also from the distribution of seismic foci. The author believes that the issue of the origin of terraces and horst-graben structures on the continental slopes is nearly similar to that of the origin of other features such as double island arcs. The appearance of the latter is often considered to result from the regular action of some concealed deep mechanism; however, it seems more plausible that double island arcs present more or less particular forms, although repeated, of the tectonic ruggedness of the

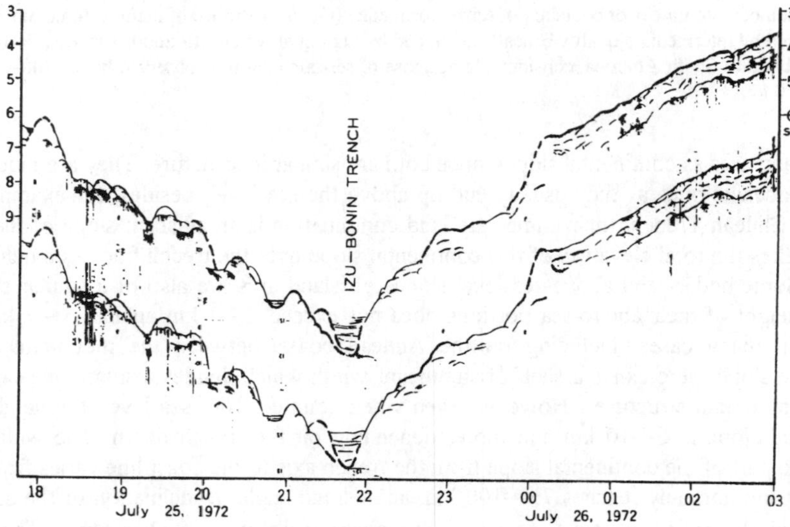

Fig. 8. Seismic reflection profile across Izu-Bonin Trench, (after Murauchi and Asanuma, 1977).

continental slope and adjacent area base where they outcrop from beneath the ocean level.

Seismofocal profiles within the part covering the base of the slope testify to the occurrence there of mainly steeply dipping faults (Figs 11, 12). They create longitudinal blocks displaced with respect to one another in a rather complicated pattern. A relative stepwise subsidence of these blocks towards the trench axis develops against a background of transverse tension which is demonstrated by numerous longitudinal grabens as well as by these steps; at the same time, individual longitudinal blocks are subject to opposite upthrusts. In this case, the number of terraces and benches cannot be determined as strictly constant. But the most general regularities of the structure of continental slopes, in their parts of different height over the trench axial valley, undoubtedly exist.

The tectonic ruggedness of oceanic slopes of the trenches is also rather complicated, but in another way, and with much lesser tectonic displacements in amplitude. The steepest feature is normally the surface immediately adjacent to the axial valley (with total inclination of up to 5°, occasionally of up to 15-20°); farther oceanward the surface is rapidly smoothed out, so a convex bend near the trench axis is formed. The foundation of the oceanic slope is dissected here by many longitudinal parallel faults which are close to one another and inclined under the surface to different sides; between them there are alternating grabens and horsts with an amplitude of vertical displacements of up to several hundreds of metres (Fig. 14), which form together a rather even general surface. According to a model made for the Hellenic Trench (Initial Reports..., 1973a), the horsts and grabens on the convex part of the oceanic slope are very traceable and, along the slope make a series of parallel swells separated by similarly extended burrows. Near the trench axis the sediments and their associated accumulative surface are often found only in the grabens being divided by outcrops of the bedrocks (Fig. 10, 24B). The sediment cover becomes continuous, or nearly so, in the higher part of the oceanic slope.

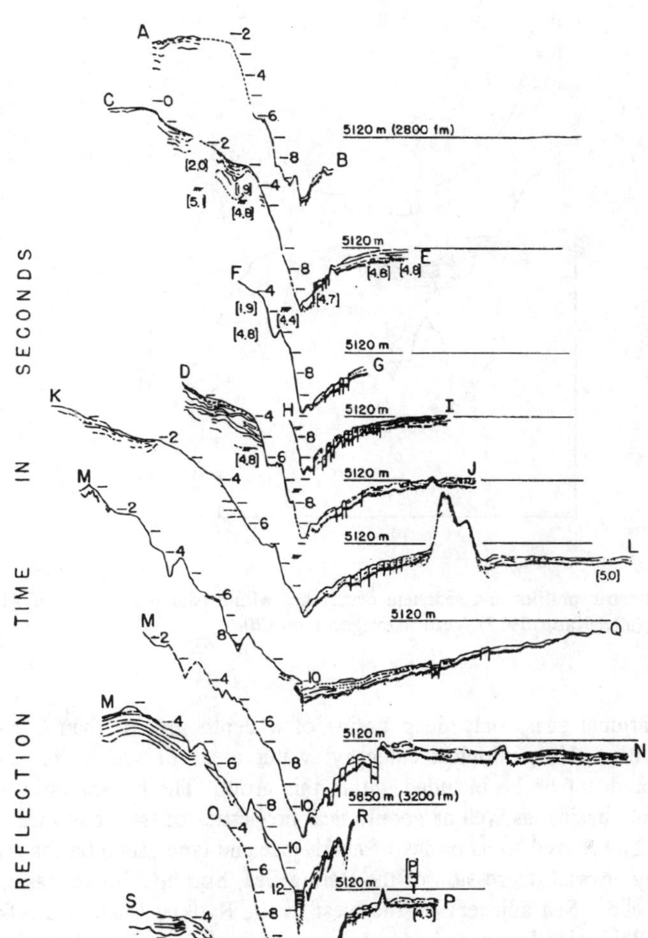

Fig. 9. Seismic reflection profiles across the Japan Trench, (after Ludwig *et al.*, 1966). Depths in fathoms and metres. Some p-wave velocities are shown in brackets.

We will not consider here in any detail the relief of island arcs since its description can be found in many works. Some considerations on the formation of island arcs or their constituent parts are expressed above and in the text below. With greater width and height of the island arc base, the linear longitudinal structures influence the forms of relief in their above-sea parts to a lesser degree; however all the island arc structures display longitudinal mountain chains and grabens dividing them.

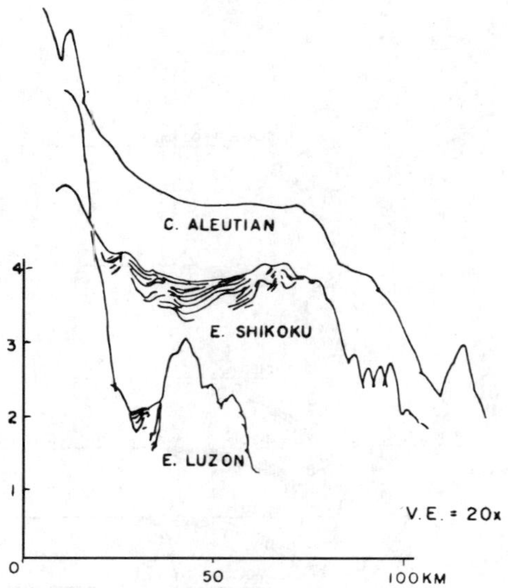

Fig. 10. Hypsometric profiles and sediment occurrence within island-arc slopes of Aleutian, Shikoku, and Luzon trenches (after Karig, 1977). Vertical exaggeration x20.

As for marginal seas, only deep basins of a depth greater than 2.5—3 km should be considered within this geological category: water areas of shelves or similar types of sea basement should not be be included within this group. The hypsometric parameters of such proper marginal basins, as well as geophysical properties of their basement, testify to the fact that they can be referred to as basins of nearly-oceanic type. Such basins are the Aleutian and Komandorskiy abyssal depression of the Bering Sea, South-Okhotsk Basin, the south-eastern part of the Yellow Sea adjacent on the west to the Ryukyu island arc (Okinawa Basin), the whole of the Philippine Basin, the deep-sea part of the South China Sea, the Sulu and Celebes Seas, and considerable parts of the Coral Sea and the Fiji Sea. The sizes of these basins vary greatly: for example, the area of the Philippine Sea is 5.7 million km^2, while that of the South-Okhotsk Basin is about 100 000 km^2; the width of the former is 2000 km and that of the latter 250 km. The greatest depth in the marginal oceanic basins varies from 6500 m in the Philippine Sea to 2700 m in the Okinawa Basin (nearly 3700 m in the Bering Sea and Sea of Japan, 3500 m in the South-Okhotsk Basin, more than 5500 m in the Sulu Sea, etc.). Among the peculiarities pertaining to the marginal oceanic basins, one should point out the presence there of the second rear deep-sea trenches (if the first trench is considered to be that on the oceanic front of the island arcs). This phenomenon is most pronounced in the Philippine Sea, whose asymmetric structure and seismofocal zones equally inclined from the ocean and accompanied with trenches on both sides have already been described above. Behind the second trench which is in front of the Philippine Archipelago, there appear also the South-China, Celebes and Sulu marginal basins.

The appearance of the second line of seismofocal zones and trenches on the inside of the marginal sea does not seem to be accidental or peculiar to only the Philippine basin. Recently buried trenches seem to have existed at the continental borders of the Bering Sea, which can be deduced from the evidence obtained there (e.g. Schmidt, 1980); but such

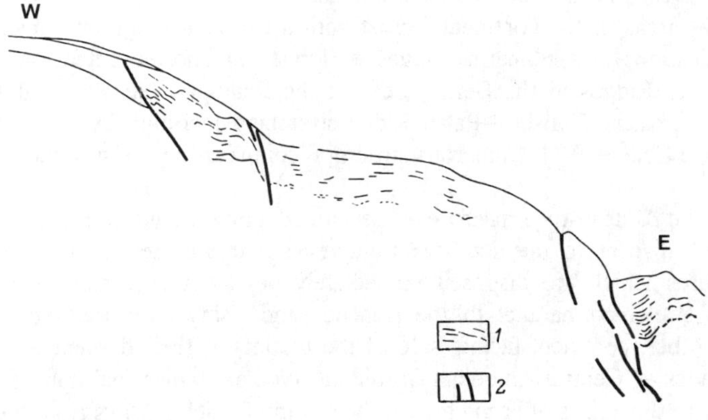

Fig. 11. Cross-section through continental slope of Japan Trench (opposite to Sanriku district) according to seismic reflection profile, (after T. Sato, 1973). 1—Reflecting plains; 2—faults.

formations were not revealed behind island arcs (Mariana and Aleutian) at their slopes facing the continents. The same may be true of the Japan and Kuril arcs. On the other hand, east of New Guinea and Australia, a second line of trenches and their corresponding seismofocal zones emerged on the continent-facing side of the island arcs (made up by the Solomon islands and a system of the New Hebrides, Fiji and Tonga Islands). The Coral marginal sea is thus restricted by a chain of deep-sea trenches as is the Philippine Sea, but in this case they are disposed not on the continental but on the island arc side of the sea.

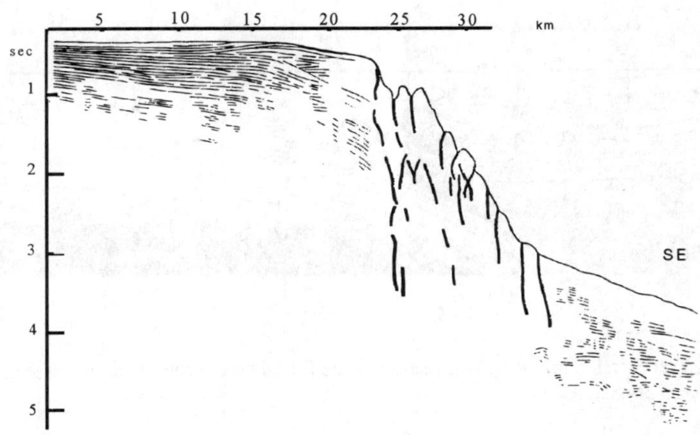

Fig. 12. Cross-section through middle part of Kamchatka continental slope near Shipunskiy Peninsula, according to seismic reflection profile, (after Gnibidenko *et al.* 1980). Depth shown in seconds. For symbols see Fig. 11.

The marginal oceanic-type seas, except for the smallest, normally have inside structural complications in the form of rises which divide them into separate basins, hollows, etc. The submarine Yamato Rise in the Sea of Japan is of an irregular shape and, according to much evidence, includes areas of the continental crust with a geophysical granite layer. Otherwise, the relief complications are submarine ridges of longer or shorter extension, such as the Shirshov and Bauer Ridges in the Bering Sea. In the Philippine Sea, a great complicating structure is the submarine Kyushu—Palau Ridge stretching meridionally through the central part of the sea (see Chapter VI). Numerous structural complications exist in the Coral Sea and Fiji Sea basins.

In the case of the Andean-type macrorelief developed along the western coast of South and Central America, adjacent to the deep-sea trenches is a continental edge with large linear mountain structures which are disposed immediately beyond the trench. These consist of mountain ridges stretching parallel to the trenches and rising over the shore for a short distance; on the other, continent-facing, side of the mountains their descent is much gentler. Ridges of the Andes — Central American Cordilleras type are a marginal uplift accompanying the trench similar to the uplifts of island arcs in their origin (which follows from the gravimetric data considered below). The difference between the system of trench — marginal continent-edge uplift from the type trench — island arc — marginal sea lies in the fact that the continent-edge uplift is not separated from the continent while the island arc is separated from it by the marginal sea. This can be considered as one piece of evidence for an opening nature of the marginal seas due to tension. A transition from one system to another and an origin of island arcs can be imagined as separation of the continent-edge uplift from the continent behind which a marginal sea opens up.

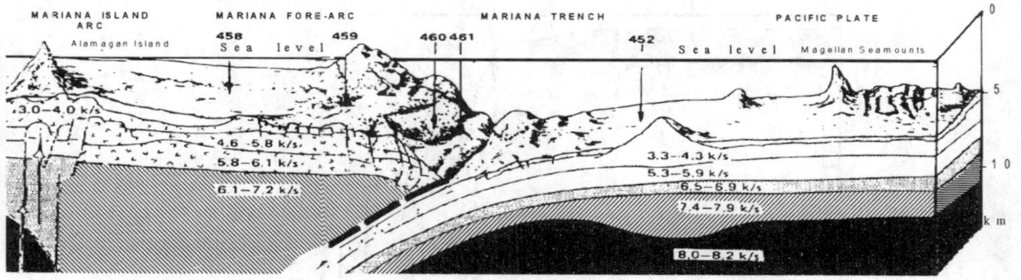

Fig. 13. Mariana Trench relief and supposed section of the lithosphere. From materials of Leg 60 by *Glomar Challenger* (*Initial Reports..., 1978*).

ORIGIN OF MARGINAL SEAS

The marginal seas were considered in the past, and are considered now by a number of researchers, as analogues of ancient geosynclinal basins of a certain type, which can also be referred to as marginal. Among various viewpoints advanced to account for the origin of marginal seas and marginal geosynclines, two can be especially mentioned. One of them attributes the origin of marginal seas to the transformation of the continental or any other crust which had existed earlier in their place, accompanied by the subsidence of the corresponding part of the Earth's surface — this may be termed a `fixation' viewpoint. The other viewpoint attributes the origin of marginal seas and geosynclines to an opening of large fault systems under which the continental crust fractures and the two parts separate from one another (so the separated massif can move horizontally from the main part of the continent towards the ocean). This moving apart process must result in the filling-up of the opening area with material of subcrust origin rising to the surface. This point of view may be called a mobilistic one (for the island arcs and marginal seas in the north-western Pacific it was developed, in particular, by A. Wegener).

There has also existed a variant of the fixation point of view according to which the marginal seas and their corresponding geosynclines were attributed to the inherited ancient primary `protogeosynclines'. A further variation of this idea consisted in the belief that the marginal seas were stringed about with island arcs that had emerged from the oceanic foundation at a certain stage of its development. Such a viewpoint should obviously be rejected now, since deep-sea drilling has revealed no sediments older than the Mesozoic Late Jurassic in either the marginal seas or over the ocean areas, which might testify to their 'protogeosyncline' development. Besides, the sedimentary cover is younger throughout the marginal seas than it is in the adjacent parts of the oceans: for example, in the marginal seas of the north-western Pacific, this cover is normally made of Cenozoic sediments, and in a number of cases only with those of the Neogene—Quaternary (the Sea of Japan), while within the adjacent ocean the sedimentary cover includes Cretaceous and Upper Jurassic sediments.

The concept of the formation of the marginal seas in the place of former continental crust was substantiated in different ways. The views suggesting that the cause is a subsidence of the Earth's surface in the frame of marginal seas due to a synclinal bowing of the crust under the action of lateral compression or other forces have now become obsolete, since it has been found that the foundation of marginal seas is more similar to the oceanic one and cannot be taken to be a subsided continental crust. The only acceptable explanation that remains is the oceanization of the continental crust which had existed in the place of a marginal sea as a result of which the continental crust turned into the oceanic one.

A hypothesis of oceanization was developed by V. V. Belousov, and his views were more than once criticized. We will not dwell on the details of the discussion during which the difficulties in physico-chemical, geochemical and thermodynamical justification of oceanization were pointed out. No less difficult is to correlate the oceanization hypothesis with geological evidence, which was pointed out by V. E. Khain, P. N. Kropotkin, Yu. A. Kosygin, V. A. Magnitskiy, V. M. Sinitsyn, W. Jacoby, and other researchers. A point that must be stressed is that both over considerable areas in the open ocean and over the marginal seas (in those parts with an ocean-type foundation), numerous deep-sea drilling holes and seismoacoustic profiling has normally established a sharp boundary between the sedimentary cover and the basement without any alteration of sediments and, let us say, the basalts, which are considered to make up a basement of the oceans and marginal seas (the volcanic interbeds which are a constituent part of the sedimentary section in marginal seas are represented throughout by tuffs arriving from the adjacent island arcs and distributed in all parts of the sediment sections). Since oceanization near the surface is usually considered to consist in outflows of basalts overflowing the oceanic floor, it should be expected to result in interbedding of basalts and sedimentary rocks, because single-act outflow of basalts in oceanization seems hardly possible. The absence of interbedding in oceanization is especially difficult to explain since, during it, no horizontal displacement is supposed which could transfer parts of the sea foundation from the zones developing intense volcanism to the zones of non-volcanic sedimentation (which take place during the spreading). A two-member section divided into a sedimentary and its underlying mafic basement parts with a sharp boundary between them is characteristic both of recent marginal seas and their ancient analogues which are called geosynclines (and marginal ones).

At the present time most researchers accept an idea of an opening origin for marginal seas. The same point of view is also adhered to by the supporters of subduction (e.g. Karig, 1971), although they experience additional difficulties in being forced to explain the distribution of unfolded sediments of marginal seas in the areas where they suppose compression to exist. However, the data obtained from deep-sea drilling and from geophysical evidence of oceanic-type basement within marginal seas suggest their expansion in spite of the above-mentioned difficulties. A mechanism which is usually proposed to account for this expansion and the formation of the oceanic foundation in the marginal seas is that of spreading. The chapters that follow, however, offer a different view on this problem.

In addition to the data immediately relating to the marginal sea basins, their opening nature is also argued for using evidence gathered from island arcs. These allow one to establish that the continental crust contributes to the make-up of the foundation in large island arcs and to compare its structure with the adjacent continental part (such a comparison can best be substantiated in the case of the Japan Arc). The existence of the continental crust with a developed geophysical granite layer in the base of a number of island arcs (Kuril, Aleutian, etc.) has contributed to formation of considerable volumes of acid volcanic outflows. Paleomagnetic evidence which is also obtained most fully for Japan, proves the continental bending in the Paleogene and Neogene sediments of this island arc towards the ocean from which an expansion of the marginal Japan Sea behind it follows (Kawai *et al.*, 1962; Otofuji amd Matasuda, 1983). There are new paleomagnetic investigations revealing a similar bending of the Kuril-Kamchatka arc (Bazhenov and Burtman, 1994). Evidence of the opening of marginal seas with the movement of island arcs away from the continents to a large degree was summarized in works by P. N. Kropotkin (Kropotkin, 1972, 1980; Kropotkin and Shakhvarstova, 1965). This idea was originally advanced by A. Wegener (1915).

GRAVITY ANOMALIES

All the areas where seismofocal zones outcrop onto the Earth's surface are associated with a complex of linear gravity anomalies. The largest ones which have no analogues in their values anywhere else on the Earth's surface are negative gravitational anomalies accompanying the axial zones of the marginal deep-sea trenches. Such an anomaly was detected for the first time in the Sunda Trench after F. Vening-Meinesz started, in 1923, carrying out a marine gravimetric survey in the area of the Indonesian Archipelago using a submarine with a specially designed gravimeter. It was established later that such anomalies are typical of all deep-sea trenches which are both in front of island arcs and in front of the American continents. There exist also two positive linear anomalies parallel to the negative one: one is on the external side of the trench along the external oceanic swell, the other on land adjacent to the trench within the island arc on the western surroundings of the Pacific and in the Andean continent-edge uplift. The land anomalies are larger and continuous, while the anomalies over the oceanic swells are intermittent in many cases.

All three gravity anomalies are especially clear in isostatic reduction, and also in the Faye reduction which, as it is known, can be considered to be isostatic when the compensation level is attributed to the Earth's centre. The negative anomalies associated with trenches are most pronounced in this case and their values reach 200—300 milligals, and sometimes more, whereas the values of isostatic or Faye positive anomalies on land or on the outer oceanic swells are of 50 to 150 milligals and up to 100 milligals respectively.

All these anomalies are displayed more strongly in the Bouguer reduction (because of its unrealistic assumption of additional or subtractional masses which everywhere as it was supposed, cause the existence of mountains or depressions on the Earth; that is why in the author's opinion the Bouguer reduction is the least feasible for the tectonic interpretation of gravity data). However, the Bouguer anomalies on the island arcs are positive, unlike the mountainous areas of most of the Earth, which indicates the closeness of isostatically non-compensated heavy masses beneath them.

It was supposed long ago, and repeatedly justified (e.g. Vening Meinesz, 1940; Artemjev, 1966), that isostatic gravity anomalies testify to large disturbances of isostatic balance. Such gravimetric anomalies uniformly built near outcrops of all seismofocal zones obviously point to uniformly directed tectonic forces acting constantly in seismofocal zones. Usually there are no disagreements between the researchers adhering to various principles with this aspect.

HEAT FLOW

The active boundaries of oceans and continents are accompanied not only by gravimetric anomalies but also by great thermal anomalies. These are expressed as distribution of high heat flow within the larger part of above-seismofocal areas. An obvious display of thermal anomalies is volcanic activity and its accompanying thermal phenomena in volcanic belts associated with seismofocal zones. The total amount of heat which is delivered to the Earth's surface in the active marginal areas is of the same order as in the mid-oceanic ridges (Magnitskiy and Artyushkov, 1978).

The two main forms of heat transfer from the depths to the Earth's surface are conductive and convectional. The latter is displayed mainly in volcanic areas while the former is the most common mechanism over the greatest part of the Earth's surface. For conductive heat, there is a heat flow parameter, which is the product of a geothermal gradient value and the heat conductivity coefficient of the rocks lying below the boundary of seasonal changes in ground temperatures. This heat flow parameter is used as the main criterion of the geothermal regime in the Earth's surface. It is measured in heat flow units (hfu) which are equivalent to 10^{-6} cal/cm^{2}sec (sometimes also by milliwatt/m^2 which is equivalent to 0.0239 hfu — but the latter unit is more convenient). The average values of heat flow on the Earth are 1.1—1.4 hfu.

The heat flow in above-seismofocal areas has peculiar features; the principal ones may be found in sections through the Kuril arc and South Kamchatka and through Japan shown in Figs 14 and 15. In the Kuril arc, the maximum values of the heat flow exceeding 2—3, occasionally of up to 5 hfu and over, are found along the island arc and on its north-western (inside) slope. The other maximum of heat flow (Profile 1—0 in Fig. 16 crossing the Island of Iturup) which is to the north-west belongs to the central part of the South-Okhotsk deep marginal basin. In the generalized profile through South Kamchatka, the heat flow increases successively from east to west, reaching 3 hfu (an averaged value) on the eastern side of the Kamchatka

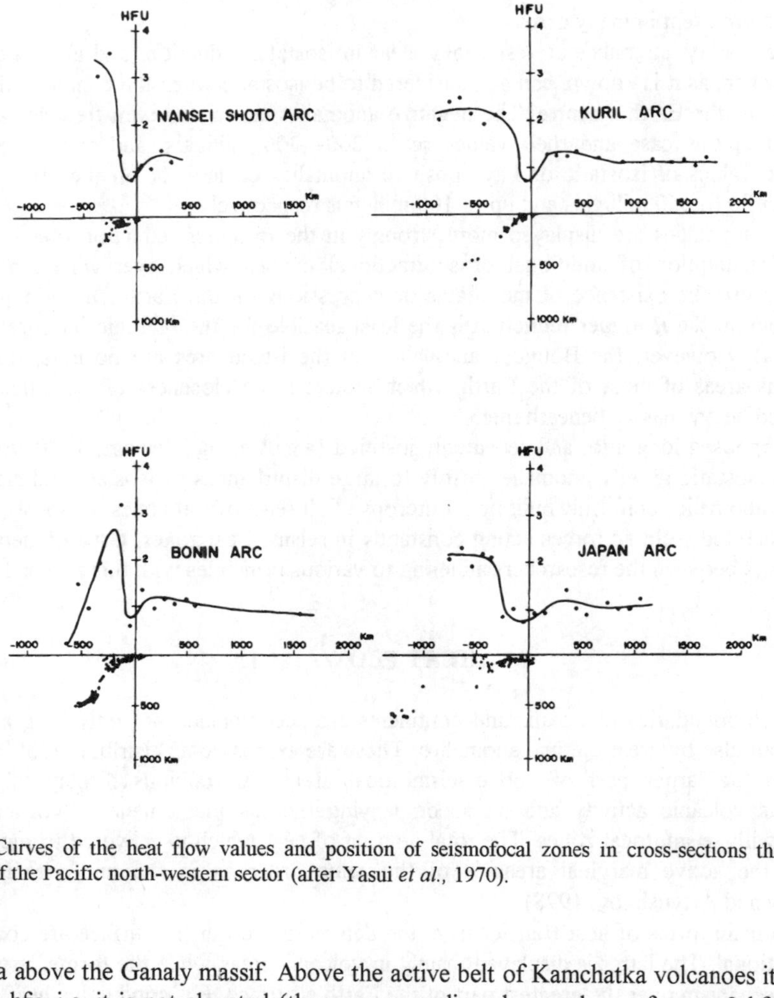

Fig. 14. Curves of the heat flow values and position of seismofocal zones in cross-sections through some trenches of the Pacific north-western sector (after Yasui *et al.*, 1970).

Peninsula above the Ganaly massif. Above the active belt of Kamchatka volcanoes it reaches a mere 1.5 hfu, i.e. it is not elevated (there are recordings known here of up to 3 hfu or more, but lower recordings nearer 1 hfu prevail). A similar comparatively low average heat flow also characterizes the zone of the Japan volcanic belt.

It may be supposed that the higher values of heat flow on the Kuril arc islands are accounted for by the fact that these islands are largely volcanic structures and the measurements there are virtually taken on the volcano slopes. The measurements on Kamchatka and in Japan substantially cover the areas at some distance away from the volcanic centres. It is probable, therefore, that the data on a relatively lower average conductive heat flow in volcanic belts should not be taken as evidence for an actual drop in supply of deep heat in the volcano zone, but conversely, indicate a large expense of thermal energy used to melt large parts of the lithosphere at depths and for other processes. These processes continue at present as is demonstrated by recent volcanic eruptions, and by effluxes of huge volumes of volcanic gases. A relative (though not ubiquitous) decrease in the conductive flow can thus be explained by a convective export of much deep heat to the surface through volcanic vents and heat expenditure at certain depths. It is hard to calculate the complete arrival of heat energy to the volcanic areas since, in addition to heat required for melting the outflowed lavas, one needs

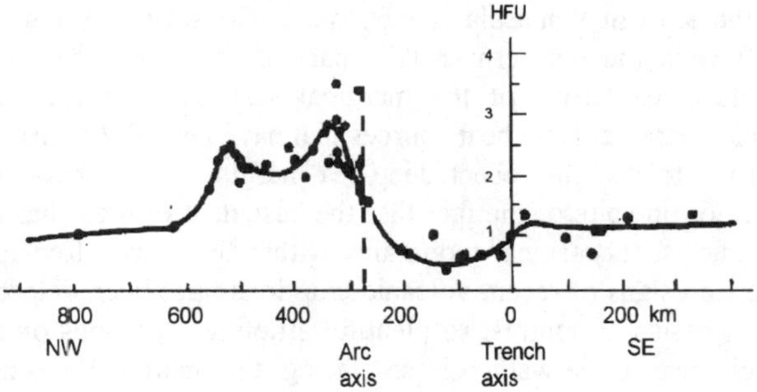

Fig. 15. Distribution of the heat flow values across Kuril-South Kamchatka island arc and Kuril-Kamchatka Trench, (after Gnibidenko *et al.*, 1980).

addition to heat required for melting the outflowed lavas, one needs to know the amount of molten magma existing at depth and to know what heat energy is transported by gases.

The above is also true of the volcanic belt on the American coast where the highest conductive flow, similar to that at Kamchatka, develops behind the volcanic belt; it takes place in a field of anomalously high heat flow, as detected in the rear of the Andean rise in Columbia.

A heightened heat flow is typical of most marginal seas. The marginal seas with higher heat flow regularly occur over the active seismofocal zones or just behind the projection of their lower boundary onto the Earth's surface; parts of the marginal seas which are before the front of seismofocal zones (i.e. in front of its associated deep-sea trench) seem to have no positive anomalies of heat flow. Thus, a higher heat flow is a feature of the South-Okhotsk Basin (up to 3 hfu) above the Kuril seismofocal zone; and of the Japan Sea, which is also entirely over seismofocal zones (2—3.5 hfu, decreasing on the Yamato Rise to 1.6 hfu); of the Okinawa Basin in the East-China Sea (behind the island arc and Ryukyu Trench to 5 hfu) of the eastern part of the Philippine Sea adjacent to the Mariana — Izu—Bonin arc (basins Parece—Vela and Shikoku to the east of the submarine Kyushu—Palau Ridge to 2.5—3 hfu); located behind the Philippine Islands, partly over the corresponding seismofocal zones, the South China, Sulu and Celebes Seas. But a heightened heat flow is absent over the West-Philippine Basin to the west of the Parese—Vela Ridge, which is situated before the front of the seismofocal zones of the Philippine and Ryukyu island arcs, and in the basins of the Coral Sea, before the front of the trenches and seismofocal zones of Solomon—New Hebrides Islands.

Behind the Tonga—Kermadec arc, a high heat flow can be found in an inter-arc basin in front of the Lau—Colvill Ridge (Lau—Havre Basin), but there is no such flow in the South-Fiji Basin located further westwards. The latter can be considered as an area in front of the seismofocal zone of New Hebrides. The deep-sea basin of the Bering Sea behind the Aleutian arc also has a positive thermal anomaly up to 3 hfu and over, although it is rather narrow, which may be due to the relatively shallow depth of the Aleutian seismofocal zone. An especially high heat flow is observed in the Komondor Basin which is a unique `hot spot' on the Earth, with a power of thermal flow exceeding that of Iceland (Smirnov *et al.*, 1982, p. 114).

Thus, the existence of a high heat flow above seismofocal zones,which is displayed equally with island arcs and marginal continental uplifts (of an Andean type), and in the deep-sea marginal basins, is quite unequivocal. The disappearance of high heat flow over the Yamato Rise, whose crust is probably mainly continental, and from the East-Korean Rise in the Sea of Japan adjacent to the Korean Peninsula is interesting. Considered along with a drop in the value of the heat flow in the northern shallow part of the Sea of Japan, this phenomenon suggests that the deep-sea basins of the marginal seas above seismofocal zones with a foundation of oceanic character have heat sources that have neared the surface.

It is important to note that the recent deep-sea marginal basins have no well-established volcanoes on their floor, in spite of the fact that the basement of these basins is considered to be basalt, i.e. volcanic; so occasional mountains within them are often taken to be extinct volcanoes. However, no signs of recent volcanic eruptions have been detected in the deep-sea parts of the marginal basins in contrast to plentiful erupting volcanoes on the adjacent island arcs (including their rear underwater slopes facing the marginal basins). As mentioned previously, deep-sea drillings in the inside parts of marginal basins have detected no signs of repeated basaltic or other lava flows as interlayers or sills among the sedimentary sections. Obviously the existing heat flow in the marginal seas is not large enough for wide volcanic activity within them (at any rate, in recent conditions), contrary to island arcs, where molten magma is constantly finding its way to the Earth's surface at many locations. The reason for these differences can be attributed to different amounts of heat delivered from the seismofocal zone levels and to the composition of the lithosphere beneath the island arcs and marginal seas. It may also be possible that the water cover which is in the marginal seas contributes to dispersion of part of the thermal energy arriving from the depths.

Numerous measurements taken in other areas (see Figs 14 and 15) have established that the value of the heat flow sharply drops with a transfer from a volcanic belt to a continental or island arc slope over the trenches. In the upper and middle parts of the continental and island arc slopes, the measured values of the heat flow are very unsteady; for example, on the slope of the Kuril-Kamchatka Trench it varies from 0.4 to 3.7 hfu with prevailing values of 1—1.5 hfu (Smirnov and Sugrobov, 1975, 1980). Towards the foot, the variations become lower, although here, too, along with figures from 1 hfu and lower one can find values of 2—2.5 hfu.

The lowest values of heat flow relate to the axial zones of the trenches, which has been pointed out by many authors. Rather common here are values lower than 1 and down to 0.4 hfu and even less (although rather seldom). It is notable that at two stations in the Kuril—Kamchatka Trench, negative values of heat flow were obtained (Smirnov, 1980; Smirnov and Sugrobov, 1979, 1980). However, in the same Kuril-Kamchatka Trench, the values of heat flow at some stations reach about 2 hfu (Smirnov and Sugrobov, 1979, 1980). There are measurements over 2 hfu in the axial part of the Nansey Trench against Shikoku Island, and at the beginning of its oceanic slope they amount to as much as about 6 hfu (e.g. Lu *et al.*, 1981). At the northern extremity of the Kuril-Kamchatka Trench and the western Aleutian extremity the values of heat flow are rather elevated to reach 2-2.5 hfu (Smirnov *et al.*, 1982), so the thermal background in the Aleutian Trench is higher than in other trenches.

Based on the inhomogeneity of the heat flow observed in the trenches with considerably different values at closely spaced stations, Ya. B. Smirnov and V. M. Sugrobov arrived at the conclusion that the measurements made in the trenches often prove to be lower than they should be due to a circulation of sea water trapped within the fractured bedrocks overlain with landslides and quickly accumulating sediments of suspension flows (Smirnov and Sugrobov, 1979, 1980). The actual heat flow measured in the trenches may be understated by as much as 10-50%, according to Smirnov (1980); therefore, a negative anomaly of heat flow in teconic trenches `may prove false', Smirnov and his colleagues have suggested (Smirnov, 1980, pp.53, 123) without any connection with the eduction idea. Most researchers estimate an average heat flow (without a correction for cooling of rocks by water circulation) is 0.9—1 hfu, which is close to the lower limit of the Earth's average heat flow and equal to an ordinary heat flow at the foot of slopes of spreading ridges.

According to S. Uyeda (1974), and some other authors, the axis of the lowest values of a heat flow, similar to the axes of negative gravitational anomalies, is shifted from the trench axes towards the foot of the continental slopes.

The value of the heat flow then rises from the axis zone of the trench towards the outer oceanic swell. Thus, at the Zenkevich swell opposite the Kuril—Kamchatka Trench it measures 1—1.5 hfu rising occasionally up to 2 or over and dropping down to 0.5 hfu (Gnibidenko *et. al.*, 1980). An average value of a heat flow over the oceanic swell is statistically estimated at 1.2±0.3 hfu. The values of 1—1.5 hfu also prevail on the outer slope of the trench between the axis zone and the oceanic swell.

SEDIMENTS AND BASEMENT WITHIN THE TRENCHES

Numerous seismoacoustic profiles allow an assessment of occurrence and thickness of sediments on the trench slopes and floor.

The sediments on the continental island arc's slopes form a cover of considerable but variable thickness. On their steepest parts and fragments, mainly near the foot of the slopes, there are occasionally no sediments. Over the rest of the continental slope there are developed sediments of thickness 1 km or more, and may reach 2—3.5 km in the `pockets' (mainly grabens, for example, on the Kamchatka slope). The surface of the above-mentioned submarine terraces on the continental slope is often of an accumulative origin and corresponds to the sediment-filled grabens. One or two stratigraphic nonconformities can often be found inside the sections of large thickness which fill the grabens, as can be seen on the acoustic profiles, while a near-to-flat pitch of sediments is retained. Inside the grabens, one can occasionally find some larger folded dislocations which, judging from the preservation of subhorizontal bedding in the near-surface higher layers, appeared under submarine consedimental conditions. Over the gently sloping parts (terraces) of the slope, conditioned by the basement escarpments, the sedimentary cover is often a replica of the basement surface, i.e. it has an overlapping character. According to deep-sea drilling results, the sediments of the continental slope are represented by muds of different composition, silts, also poorly sorted turbidite sands and partly by gravel material, occasionally fine-pebbled conglomerates. According to dragging evidence, the bedrocks over the areas with no sedimentary cover consist mainly of varieties distributed in the outcrops of the adjacent land; they are the various volcanites, intrusive rocks, including those of acid composition, sedimentary rocks metamorphosed to differing degrees, and metamorphic rocks proper, all of them widely varying in their age. Dragging exploration near the feet of continental slopes has detected metamorphic schists of a greenstone type, amphibolites and serpentinites. The outcrops of bedrocks often bear a patina of submarine weathering.

The sediments on the trench floor, in the region immediately beyond the foot of the continental slope (i.e. in the axial part of the trench proper), differ considerably — first of all, in thickness (Figs 16 and 17). With horizontal and subhorizontal bedding, their thickness is seldom over 250 m, and occasionally much less. Furthermore, near the axial part of the trench there are often areas of outcrops of bedrocks (seismoacoustic foundation). Because of the great depths involved, this material can seldom be raised to the surface; but in its geophysical properties, it usually corresponds to basic and ultrabasic rocks. In the places where they can be dragged clear (mostly already on the outer slopes of the trenches), they are represented by

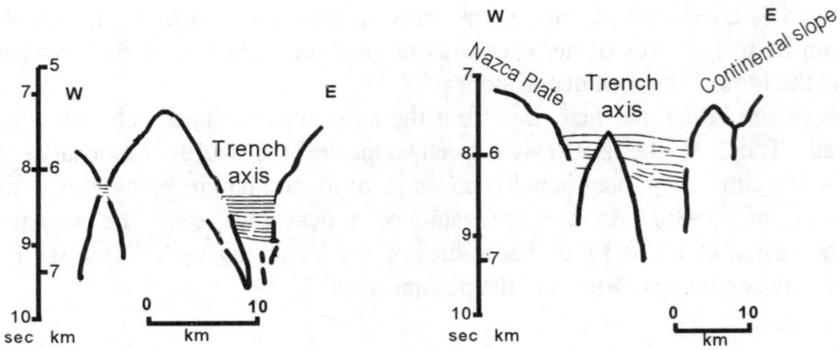

Fig. 16. Occurrence of sediments in the Peruvian Trench, according to seismic reflection profiling, (after Prince and Kulm, 1975).

greatly altered rocks which are usually taken for paleotypic basalts and their tuffs, and by the metamorphosed gabbroides and serpentinites. In essence, sediments in the axial valley of the trenches, similar to the continental slopes, when they are of small thickness, occur in the grabens which complicate the relief of the basic foundation. This is seen, for example, over the profiles of the Peruvian—Chilean Trench (Figs 16 and 23).

The above is true of the deep and large trenches which accompany the deep and pronounced seismofocal zones — Kuril—Kamchatka, Japan, Mariana, Tonga, Peruvian-Chilean trenches. In some of the shallower trenches that accompany fragmentary seismofocal zones and which do not reach maximum depths, the thickness of sediments in the axial part may be greater. One example for this is the Hellenic Trench in the Mediterranean Sea whose sediments in the axial part, according to evidence from seismic sections, reach up to 1 km; bores from deep-sea drilling revealed Quaternary sediments of up to 480 m in thickness (*Initial Reports, 1973a*). As in other trenches, those sediments occur horizontally. There are also trenches fully buried with sediments and only discoverable during seismic explorations; in such cases, there are no signs of the existence of recent seismofocal zones. Buried trenches of such a type are found, for example, near both coasts of North America (Emery, 1965; Scholl, 1974; Sheridan, 1974). These sediments also occur mainly as horizontal and subhorizontal. The thinnest sediments on the trench floor seem to correspond to the most active seismofocal zones. In other words, the trenches become filled up with sediments as the activity of seismofocal zones fades.

In their composition, the sediments of trenches are rather similar to the sediments of the continental slopes. Besides abyssal muds and clays, there occur here turbidites containing interbeds from coarse-grained to gravel and small-pebbled (according to dragging evidence). Some gentle stratigraphic nonconformities can be seen in seismoacoustic profiles in the graben structures of the trench floor (see Figs 16 and 23), and the supporters of subduction sometimes suppose such discontinuities confirm the plate movements and reflect deformations caused by plate underthrusting (e.g. Prince and Kulm, 1975). But such an opinion cannot be accepted since the nonconformities are too negligible; after the seismic sections are brought to a normal ratio of the horizontal and vertical scales, their angular value virtually defies determination. Similar, and greater, nonconformities are common in the sediments deposited by currents and by turbidite flows too.

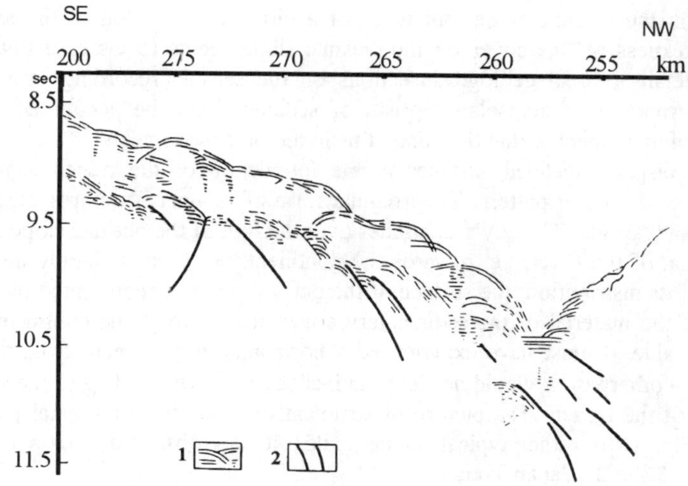

Fig. 17. Section over axial part of Kuril-Kamchatka Trench near southern end of Kamchatka. Based on seismic reflection profiling, (Gnibidenko *et al.*, 1980). 1—Reflecting planes, 2—faults in acoustic foundation.

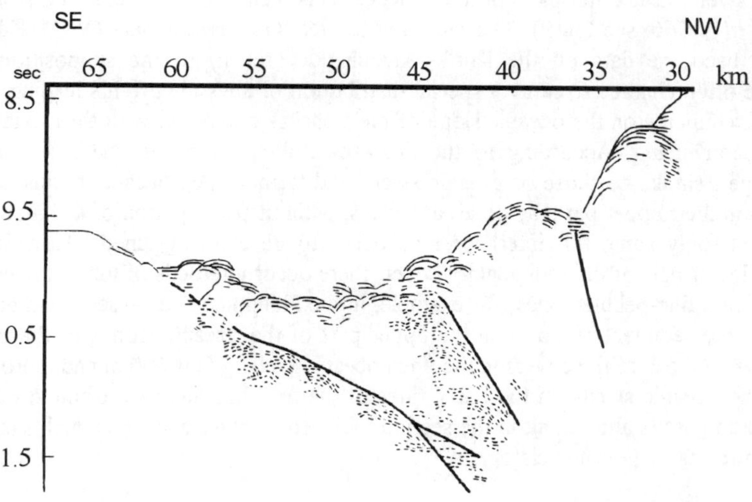

Fig. 18. Section over axial part of Kuril-Kamchatka Trench near its northern end. Based on seismic reflection profiling, (Gnibidenko *et al.*, 1980). For symbols see Fig. 17.

The sections obtained in seismoacoustic profiling also display other forms of distribution of sediments in the axial part of the trenches, in addition to their concentration in grabens. Fig. 16 shows an instance when the axial slit of the trench accommodates a thin sedimentary prism of a triangular cross-section with horizontal bedding. Just below, in the bedrock foundation, the inclined tectonic surfaces start and submerge further under the continental slope; in their orientation, they continue the oceanic slope surface and probably constitute a zone of main tectonic movements. A seismically transparent layer corresponding to the sedimentary cover can also be seen on the oceanic slope, but it is not a direct continuation of the axial zone sediments. The thickness of this cover on the oceanic slope seems to increase towards the ocean (although details of real geological sections on the seismic recordings are liable to distortion). The formation of an isolated prism of sediments can be accounted for by a temporary halting of movement within the zone of main tectonic movements.

Figure 18 also displays inclined surfaces in the foundation of the trench axis but the sedimentary cover is of another pattern. The irregular form of its inner structures suggests that it is of mainly landslide origin. The cover continues over the foot of the oceanic slope taking up an inclination typical of the latter, i.e. towards the continent, but then is sharply interrupted. Behind the front of its distribution, the surface of the oceanic slope is represented by outcrops of the bedrocks. If the material of the sedimentary cover arrived from the continental slope, which is most probable, it must have experienced a horizontal displacement from the trench axis to the ocean — otherwise it would not have an inclination corresponding to the surface of the oceanic slope of the trench. The picture of sediment distribution in the axial part of the trench shown in Fig. 18 is rather typical; in particular, it was observed over a number of seismoacoustic profiles in the Japan Trench.

The sediments of the oceanic slope of the deepest trenches, as already mentioned, form a less discontinuous cover compared with the near-axis part of the trench, and the outcrops of the bedrocks (acoustic foundation) are the less frequent, the nearer to the outer swell. The sedimentary cover in the open ocean is known to be continuous. Furthermore, the seismic profiles reveal well the increase in thickness of the sedimentary cover from the trench axis to the oceanic swell of the outer parts of the trenches, as is seen, for example, in the Japan Trench (Ludwig *et al.*, 1966; see Fig 9). This situation has led some researchers (V. V. Zdorovenin, 1972, who used the data on the Kuril—Kamchatka Trench) to the supposition that the trenches are often characterized by a special distribution of abyssal currents leading to greater amounts of sediments on the oceanic slope of the trenches compared with their axial part, but this is not convincing. According to the deep-sea drilling evidence, the sediments of the oceanic slopes, similar to those of oceanic swells and farther open oceanic basins, are always represented in their upper parts by oozes and clays, without participation of a coarse turbidite-type material (only some tuff interbeds may occur to be coarse-grained). However, in the middle and lower part of the sedimentary cover, there occur interbeds of turbidite character up to and including fine-pebbled ones (for example, in the Philippine Sea — see Chapter VI). The thickness of the sedimentary cover in the upper part of the oceanic slopes of the trenches is rather similar to those of the covers in the open ocean, reaching 500-600 m and more.

Within the oceanic swells on the outer sides of the trenches there were found outcrops of non-weathered basalts alternating in the section with sedimentary deposits which suggests that these structures had a volcanic history.

Chapter II. Critical Overview of the Subduction Concept

DEVELOPMENT OF SUBDUCTION VIEWS

In the early 1960s, R. S. Dietz expressed an opinion on expansion of the oceanic floor from the mid-ocean ridges (sea-floor spreading) and subthrusting of the ocean crust under the continents (Dietz, 1961). This was an idea that had been proposed earlier by H. Hess (Dietz, 1968). They proceeded from a well-known notion on the existence of mantle convection cells. According to their suggestion, the mid-ocean ridges match the zones of divergence of convection cells while the opposite parts of those cells have a descending direction of mantle currents under the neighbouring continents. As a result, the ocean crust, which is a mantle that has been elevated onto the surface, is pushed under the continents, as it is supposed, in the marginal oceanic trenches. `Since the model follows from a concept, the author does not attempt at defending it' (Dietz, 1962, p. 289). This author believes that the underthrusting is caused by resistance which continents set up to a spreading mantle. According to such a concept, the edges of each ocean must have its borders for underthrusting. The fact that there are no such borders in the Atlantic or Arctic Oceans or in most of the Indian Ocean, was explained by Dietz by a drift of the adjacent continents together with the sima. However, this general explanation is obviously inapplicable to, for example, Africa which cannot drift in the above-mentioned way to the west or to the east since it is surrounded on three sides by mid-ocean ridges. Besides, inside Africa there is a submeridional rift which matches, according to Dietz, the place of encounter of the ascending flows of the mantle cells.

In Dietz's construction, the basic idea proved fruitful concerning spreading of the oceanic floor from the mid-ocean ridges. Later it was reinforced with numerous data from deep-sea drilling and geophysical studies evidencing the spreading of the symmetrically disposed stripe magnetic anomalies on both sides of the middle ridges. The description of these anomalies together with their interpretation in the context of spreading was first made by F. Vine and D. Matthews (1963).

However, our task is to consider the idea of subduction. The notion of the disappearance of excess lithosphere generated by spreading in mid-oceanic ridges by means of their underthrusting under the continents was further developed in a series of papers which were published in 1968 and marked the beginning of the new global tectonics. The first to appear was the paper by W. J. Morgan (1968), in which the author suggested a geometrical model to describe a modern process of continental drift. In this model, the Earth's surface was divided into several large blocks which unite the continents with parts of the oceans; they were supplemented with some smaller blocks of similar character. The uniting of the continents and oceans into single blocks, later called `plates', made a new contribution to the concept which was being created. The plates were suggested to be `absolutely' rigid, both in their continental and oceanic parts; they were to be built up with new oceanic crust on the side of the mid-ocean ridges — on their flank slide relative to each other — while on the opposite to the mid-ocean ridges side they must be represented, in the typical case, with boundaries of a trench type (also called compression boundaries), where the plates were `consumed'.

The fact that the author did not acknowledge the opposing of the oceanic and continental crust and failed to recognize the difference in their mechanical properties allowed him to transfer the boundaries of destruction from one ocean through the adjacent continent to the other. This allowed, in turn, the connection of movements in the remote mid-ocean ridges and trenches. However, the quantity of boundaries represented with trenches turned out to be insufficient, and so they were supplemented with boundaries formed by some submarine but not mid-oceanic ridges and by other formations, which was anomalous in Morgan's opinion. Taken as an example of the latter was Macquarie Ridge which extends south of New Zealand. `We suppose this ridge is the result of slow compression; and that fast compression leads to the trench-type structures' (Morgan, 1968, p. 1960). The rather small Macquarie Ridge (1200 km in length) remains a unique example of such a ridge; the mentioning of it is followed with the note: `We have supposed that slow compressive systems are difficult to identify and have freely placed such boundaries in likely places' (*idem*). In this way, `a compressive boundary has been placed in the Mediterranean Sea between Europe and Africa.' (*idem*); the author also mentions `the compressive-type features'of the mountains in Iran and Turkey (*idem*, pp. 1959-1960). These remarks, quoted quite fully, are what was said about the processes occurring on compression-type boundaries, in cluding those which are accompanied by deep-sea trenches. The rest of Morgan's paper is devoted to a description of the supposed displacements of the plates.

A somewhat broader consideration of the subduction concept is given in the paper by X. Le Pichon (1968). As in Morgan's work, it is mainly devoted to the issues of spreading and plate movements (Fig.20). Having adduced grounds for notions of spreading, mainly in conformity with the views of Vine and Matthews (1963), Le Pichon raises quite a reasonable question of whether the Earth as a whole expands with the expansion of the oceans in consequence of spreading? The proof of no expansion of the Earth advanced by Le Pichon to support the new global tectonics is as follows (proof is unique, not only in this paper but probably in the literature devoted to plate tectonics, excluding some interpretation of paleomagnetic data, which will be discussed below). The author points up the intersection of different numbers of mid-ocean ridges by variously orientated great circles on the Earth's sphere. So, the equatorial circle intersects the Atlantic, Indian, and East-Pacific mid-ocean ridges whose total expansion in the direction transverse to their strike is about 17 cm per year; but the longitudinal great circles intersect the Circum-Antarctic Ridge only, and it gives their expansion calculated in the same manner between 0 and 7 cm/yr. `There is no evidence',— Le Pichon writes,—`for important pole migration during the last 10 m.y., yet the equator... would have increased its circumference by 1700 km and some of the great circles of longitude would not have expanded. This implies a differential expansion of as much as 270 km between the average radius along extreme great-circle circumferences and possibly as much as 500 km between individual radii. This is unacceptable, and it becomes even more so when we recognize that the present pattern of spreading has prevailed during the whole Cenozoic era. Consequently, in the expansion hypothesis, we have to assume some compensating large-scale processes of Earth's surface shortening by compression or thrust to maintain the nearly spherical surface of the Earth. The expansion hypothesis then lost most of its appeal' (Le Pichon, 1968, p. 3674). However, this argument is incorrect because it ignores an increase of the mid-ocean ridges also along their axes in the course of the opening of the spreading. This fact is obvious from a comparison of the length of the parts of mid-ocean ridges (if they are the traces of the continents' breakup) with the length of their corresponding continental coasts: the former are always 1.5—1.8 times longer (for example, such a ratio exists between lengths of the middle and south part of the Mid-Atlantic Ridge and those parts of the African and South-American coasts corresponding to its configuration; also between the Circum-Antarctic Ridge and the corresponding part of the Antarctic coast; and so on). So the oceans of spreading origin expand in the process of spreading in different (all) directions, but not only transverse to the crests of the spreading ridges, which direction is only most convenient for the analysis of stripe magnetic anomalies. This fact undoubtedly does not correspond to the model of rigid plates and is ignored by supporters of this model. The cited quotation contains also the main grounds of the subduction idea: it is generated from an unproven postulate of constant Earth size. These should be taken into account: the proofs of subduction proposed by the architects of plate tectonics were, from the beginning, and in contrast to spreading, purely theoretical.

The `processes of Earth's surface shortening by thrust' in the trench areas are those of subduction, but in the cited work this term was not yet used (although it had appeared in tectonics earlier, though in a less specific sense). Le Pichon used for the trench zones the term `sinks', and so did other authors (Isacks *et al.*, 1968). The process of subduction was designated by them as underthrusting or consuming; the word `subduction' itself came into use as a plate tectonics' term some time later (e.g. Coleman, 1971). Now, seismofocal zones at the ocean margins are not only conceived but described always as subduction zones in many works. However, because exact knowledge of the tectonic process which creates these zones has not yet been clarified, such a name should not be used.

The disposition of the subduction (`sink') zones was determined by Le Pichon, as by Morgan, rather approximately: `Although the axis of a spreading ridge is a well-defined block boundary, the location of the other boundaries (i.e. the subduction ones—Yu. Ch.) are not as well defined. The choice is guided by the seismic and physiographic data. Active deep-sea trenches and the Alpine—Himalayan mountain belt are obvious choices' (Le Pichon, 1968, p. 3676). So, proceeding from the conclusion on the constant size of the Earth and an inference of an indispensable existence of plate destruction and the disappearance of boundaries, the latter was selected as the most suitable among the known formations of the Earth; these are the trenches, where there are enough of them, and mountain ridges.

The next paper after Le Pichon's was one by B. Isacks, J. Oliver and L. Sykes (1968). The seismologist authors adopted the scheme of plates proposed by Le Pichon considering it to be correct because the basic plate boundaries—mid-ocean ridges, trench zones and the Alpine—Himalayan belt — match the world's main seismic belts (although, as the author of this book should note, with an obvious exception of the boundary between the North American and Eurasian Plates which is not accompanied by any distinct seismic belt; the question about the very existence of the plates is not discussed here since it is considered in Chapter VII, but it should still be pointed out incidentally that the disposition of growing and consuming borders, i.e. of the mid-ocean ridges and trenches, within the six main plates identified by Le Pichon does not agree with the main idea of their conveyor-like movements, and, moreover, each of the plates, when verified with the criteria suggested in the plate hypothesis, prove to be expanding).

We shall focus upon the substantiation of subduction notions proper which were expounded by Isacks and co-authors most consistently and convincingly. The works that followed added little to the statements they had made, none of them having analyzed the subduction issues as a whole in greater detail.

Isacks, Oliver and Sykes introduced the term `new global tectonics' to designate the supported concept, a term which was later to become widely used. They pointed out that their work was devoted to upholding the new concept and therefore could not be unbiased. This remark testifies to their striving to be impartial; yet some arguments suggested in the paper do not appear so. So in the authors' scheme of the slip vectors derived from the accepted earthquake mechanism the arrows show the horizontal components in directions of relative motions of the plates. They believe that the scheme demonstrates the coincidence of the vectors obtained from seismological data with the vectors of the plate motions in the scheme by Le Pichon (Figs 19, 20); however, the possible relative displacements should have been shown with two oppositely-directed arrows, since neither of the wings of the slip planes in an earthquake focus have grounds to be considered as fixed in some rigid coordinates. If one confines oneself to a single vector, there would be equal grounds to show the arrows of the

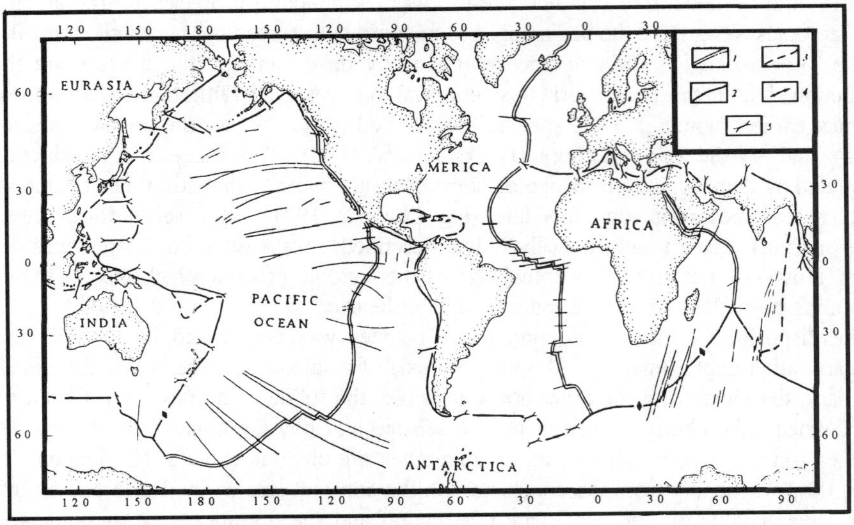

Fig 19. First plate model by Le Pichon (1968). 1—Boundaries of extension and growth of main six plates (mid-oceanic ridges); their rates were determined from time scale of stripe magnetic anomalies; 2—boundaries of those plates where the rates of their movement were computed in other ways; 3—boundaries of other possible plates; 4—symbols of extension on plates' boundaries where rates were determined indirectly from interpretation of stripe magnetic anomalies; 5—symbols of compression and underthrusting of main plates at their subduction boundaries, where their rates were determined from values of extension of mid-oceanic ridge (i. e. by repeating of these values from opposite sides of plates). Length of symbols 4-5 in legend shows scale of rate of expansion and underthrust (10 cm/yr).

reverse direction and, following the author's reasoning, it would be possible to say that they contradict Le Pichon's scheme.

The authors start their consideration of subduction with the same reasoning which is used by Morgan and Le Pichon: `If crustal material is to descend into the mantle, the island arcs are suspect as sites of these sinks' (Isacks *et al.*, 1968, p. 5866). Further, with reference to Vening-Meinesz, Benioff, Hess, and Dietz, they continue: `The asymmetric structure of the arcs and the associated pattern of earthquake occurrence in the mantle led many investigators to postulate that the structures are the result of compressive stresses normal to the arc...'(*idem*). Thus, the authors also admit that the main cause for selecting trenches and island arcs as submergence and compression-undergoing zones stemmed from a necessity to find a place of escape of the excessive lithosphere which appears when mid-ocean ridges spread, which proceeds mainly from the supposed constant size of the Earth.

To account for the lithosphere submergence, Isacks and his co-authors made use of seismological evidence on the existence of seismofocal zones. The model they suggested is as follows: seismofocal zones accompany the boundary of the plates underthrusting under the others and reflect the depth of penetration and shape of the underthrusted plate. So the latter turns out to be a flat inclined sheet, occasionally such a sheet, undergoing deformation or division into several parts (Fig. 21). The thickness of the rigid lithosphere which makes up the oceanic plate is taken to be 100 km or somewhat thinner; no explanation is given for the causes which could determine the place where the plates bend and start underthrusting.

The fact that the slab preserves the shape of a flat sheet from the surrounding mantle down to the depth of 720 km is considered to result from a relatively low temperature peculiar

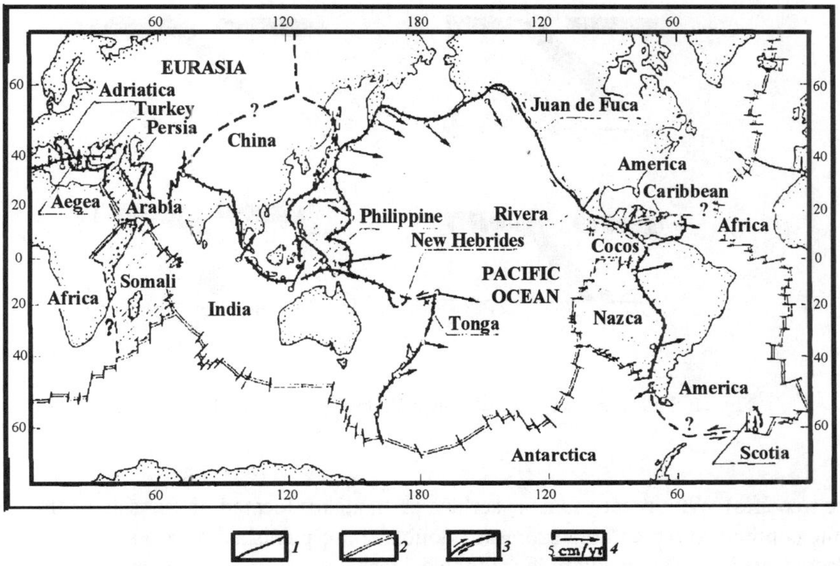

Fig. 20. Later plate model, (after Le Pichon *et al.*, 1973. 1—3 boundaries of plates: 1—subductional, 2—spreading, 3—transform; 4—relative rates of plate movements in scale, cm/yr. The six hatched plates are additionally introduced according to Morgan.

to it right down to this depth: `Judging from the recent estimations of velocities of earthquake waves propagating over the lithosphere and from other parameters, the submerging slab seems to remain much colder than its surroundings for a long time (Isacks, *et al.*, 1968, p. 5868); the evidence cited fits `the concept of a cold, relatively strong slab travelling through a rather viscous asthenosphere (*idem,* p. 5879). The orders of values of temperatures and of their changes during submergence of slabs are not shown.

Seismological data employed by Isacks and his co-authors were used as data to construct a subduction model, and the possibility of constructing such a model was to a large extent considered to be evidence for its being true. But these data contain no direct evidence for existence of a subduction process. The following seismological data were under their consideration and the conclusions they made are as follows:

(1) Seismofocal zones beneath the island arcs (i.e. to the volumes of earthquake foci distribution) are linked from below with anomalous sheet-like volumes of the upper mantle characterized by heightened velocities of the earthquake wave propagation and their low absorption (high seismic quality). The thickness of such zones (see Fig. 7) is estimated at about 100 km (but it should be pointed out, the most thorough investigations carried out in the area of the Kuril-Kamchatka arc showed that they consist of many parts of very different quality — Fedotov, 1974). So the zone of high quality was identified with the submerging slabs.

(2) Near-surface earthquakes beneath the trench axes and the oceanic slopes of the trenches were considered, according to the available solutions, to originate by tension; and grabens on the oceanic slopes were postulated to result from a bend of the underthrusting lithosphere where it begins to submerge (but the authors did not mention grabens on the continental slopes

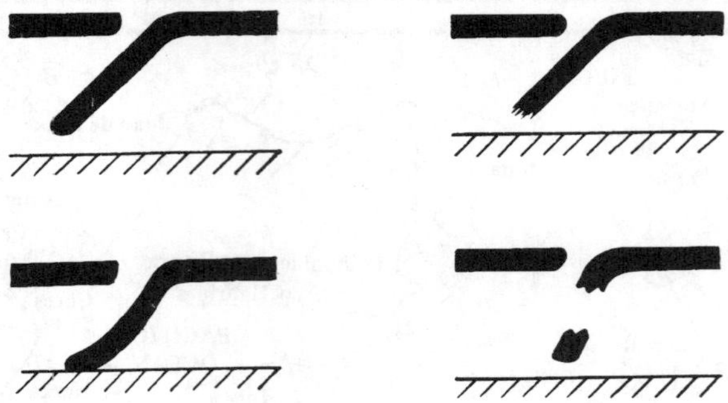

Fig. 21. Some supposed cases of plate underthrusting, (after B. Isacks *et al.*, 1968).

of the trenches). With regard to near-surface earthquakes located on the side of the island arcs and the continents, the authors expressed doubts about published data on distribution of shear displacements in their foci and preferred the evidence of overthrust-type movements.

(3) Among the earthquakes occurring deeper than 100 km, no predominance of overthrust-underthrust movements supposed to be typical of the near-surface parts was suggested; the most distinct regularity was suggested to be the orientation of axes of maximum compressive strains and, less frequently, tensile strains towards a dip of the seismofocal zones (however in a later paper by Isacks and Molnar (1971), based on more complete data, the authors admitted that the compressive and tensile strains oriented towards the dip of seismofocal zones are distributed equally). Orientation along the dip of seismofocal zones of both tension and compression strain axes leads to a single result — to displacements in the foci of the earthquakes (according to their `double-couple' mechanism which is used), along the planes inclined to the dip direction, i.e. along the critical sections of the strain ellipsoid, which means that they are nonconformable to the dip plane of the seismofocal zones. The authors point out: `The slip planes and slip directions are thus systematically nonparallel to the seismic zones; the orientations are therefore difficult to reconcile with a simple shearing parallel to the seismic zone as suggested by the common concept of the zone as a large thrust fault' (Isacks *et al.*, 1968, p. 5873). The authors reconcile the discrepancy between various orientations of the displacements in the earthquake foci and the direction of the supposed underthrusting in the following way: `the deep earthquake mechanisms reflect the stresses in the relatively strong slab of lithosphere and do not directly accommodate the shearing motions parallel to the motion of the slab as is implied by the simple fault-zone model. The shearing deformations parallel to the motion of the slab are presumably accommodated by flow or creep in the adjoining ductile parts of the mantle" (*idem*, p. 5874).

The model by Isacks and co-authors seems to contain obvious contradictions, partly noted by themselves. First of all, earthquakes prove to be a direct consequence of slab displacement, but occur not on its boundaries but only inside it, due to the emerging strains; thus, the earthquakes are generated by secondary internal deformations in the slab while the main process of the slab penetration into the mantle is not accompanied with rupture deformations and generation of seismic energy: they turn out to be forbidden. Secondly, one can also see a contradiction between an assertion of the paper (*idem*, p. 5875) according to which

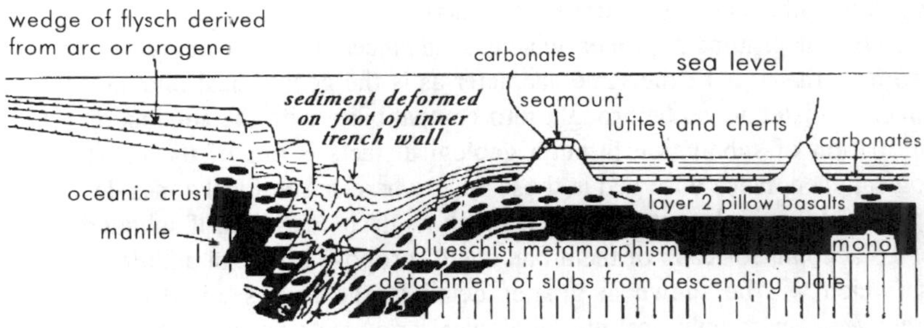

Fig. 22. Subduction at western margin of South-American continent, acc. to J. Dewey and J. Bird, 1968. Trenches are filled by deformed sediments.

`the overriding plate of lithosphere is in 'contact' with the down-going plate along the seismic zone', and a conclusion on distribution of earthquake foci only within the underthrusting slab. On the whole, this seminal work for the subduction hypothesis contains only disputable explanation of seismic phenomena from a subduction point of view, but not the seismological vindication of the latter.

A number of contradictions in the subduction model are linked with the question of the temperature state of the subducting plate: these will be dealt with below. It is also worth considering the exposition of the causes of plate submergence. The advocates of plate tectonics indicate two causes — drawing of the plates with down-going flows of the mantle currents within the convection cells (Dietz, Morgan) and their submergence due to gravity because their density is greater than that in the surrounding mantle (Le Pichon; Isacks, Oliver and Sykes). The explanation of plate submergence by means of the mantle flows appeals to its supporters because this allows the submergence process to be considered as forced, independently of the physical properties of the plate. However, in this case, one has to use a scheme of convection cells with its justly-criticized artificiality in their shapes, and also to confront the contradiction between the closed convection cells leading to mantle mixing and the data on the presence of the density and other mantle stratification. The convection scheme is also difficult to correlate with the real Earth's surface forms as well as with the supposition of rotation of plates around numerous poles scattered all over the Earth's surface. As Le Pichon (1968) points out, `it is difficult to imagine large-scale convection currents rising immediately below the crests of the ridges, with elongated conical shapes on the ridges, rates of motion increasing from zero near the axis of rotation, and staggered sections offset along transform faults' (*idem,* p. 3673). The idea of a gravitational submergence of slabs, however, results in no fewer difficulties, since a slab of essentially basaltic composition has to submerge into the ultrabasic (and consequently denser) mantle. The suggestion that the slab submerges due to its cold state, which increases its density, is difficult to accept even because when it was on the Earth's surface as the ocean foundation and was the coolest, it did not sink in the mantle.

The above covers all the basic arguments advanced in favour of the subduction hypothesis when it was evolved. These comprise mainly geophysical, and mostly seismological, data providing evidence of a mere indirect nature concerning the direction of tectonic movements in seismofocal zones. The idea of subduction proper emerged not by observations, but by attempts to correlate the evidence on the expansion of the oceans with the idea of constant size of the Earth (in the case of Dietz, this was also connected with reflective consideration of the morphology of hypothetical closed convection cells).

As to geological argument proper which is advanced to back up plate tectonics, it is necessary to note that it is of the same character as is the geophysical one, i.e. it is aimed to show that factual material can be brought into compatibility with the subduction scheme. Any direct confirmation of subduction through geological facts is out of the question, since the basic geological phenomena observed in those areas where seismofocal zones develop prove to be quite opposite in their character to those which might be expected if subduction did in fact take place. To account for each of them one has to propose a special additional complicating hypothesis (which will be shown in greater detail in the next section). Overall, the basic difference between the geophysical and geological arguments used to support subduction is that the former makes use of the total scope of modern information whereas the latter employs only the most general part of it, often reduced to simplistic abstractions. All the vast concrete geological material related to the history of development of active marginal areas and to the various and thoroughly revealed structures distributed in them are never used completely enough. This is not a mere accident — this circumstance reflects the impossibility of an explanation in principle of most geological facts from the subduction point of view. For example, observations were conducted where a marginal sea had existed in the geological past, and they testified to its consecutive opening; the subduction idea would call for these observations being compatible with an underthrusting of the sea foundation under the adjacent continent. When considering small-scale abstract schemes, one can accept suggestions of sudden turns of flows in the mantle, and so on; but the data relating to the actual features of development of the Earth's surface cannot be adjusted to provide an understanding of geological events in this way. Hence the attempts to apply subduction concepts to actual data from regional geological investigations result in disappointments, or else are not made at all.

Examples of geological development of subduction concepts were offered by the well-known works by Dewey and Bird (1970a,b) and Dewey *et al.*,(1973). Such speculative sections as in Fig. 22 (*idem,* 1970) are frequently reproduced in other works and accompanied with extremely simplified small-scale schemes instead of maps. Unfortunately, they are usually most detailed when they relate to past geological ages or to modern great depths.

DIFFICULTIES IN EXPLAINING THE SUBDUCTION PROCESS

Although the subduction viewpoint is often combined together with a much better substantiated idea of sea-floor spreading, and thus proposed as part of a developed theory, this viewpoint, when considered individually, cannot be treated other than as a hypothesis whose practical application leads to great impediments. We shall consider the most conspicuous of them.

Irruption of oceanic lithosphere into the mantle. According to the subduction hypothesis, hard oceanic lithosphere intrudes into the hard mantle, the lithosphere being of the form of a non-deformed slab 70—100 km thick and up to 1000 km in extension along the dip, i.e. 10 times as large. The displacement of the slab thus proves similar to the penetration of a hard sheet into a low-viscous medium. The evaluation of the viscosity of the Earth's crust and upper mantle do not provide grounds to consider it to be possible. With a value for the Earth's crust viscosity of 10^{22} poises, the viscosity of the prevailing parts of the upper mantle is estimated at no less that 10^{21} poises (and viscosity of the asthenosphere, which is a subordinate part of the upper mantle, is 10^{20}—10^{19} poises, which is greater by two or three orders than the viscosity of steel under surface conditions). Interaction of solid masses which are similar to one another in their mechanical properties cannot but be accompanied by their mutual deformation, i.e. with the submerging slab losing its shape even at the beginning of its way into the mantle.

Bending and change of plate shape. Where the underthrusting plate ceases to be a horizontal basement of the ocean and begins to plunge, it experiences a sharp bend. Its line corresponds to the intersection of the convex surface of the Earth's sphere with a concave surface (according to the shape of most trenches) where the plate is subducting.

The transition of a plate 100 km in thickness from a convex shape to concave (or flat) cannot occur without it being destroyed and ceasing to be a single body. Since the entire plate passes successively over the bend line, its submerging portion cannot preserve a definite shape, in particular, that of a slab. This was pointed out by Ye. N. Lyustikh (1975). He also noted that simple warping of a flat plate 70 km in thickness results in destruction of its larger part too. `With a curvature radius of the bottom surface of a plate equal to 200 km, its upper surface stretches while the lower one contracts with respect to an average surface by 15% or more. Meanwhile, the lithosphere material cannot survive deformation of more than 1—2% without losing its strength. This means that the strength can only be preserved by a 5—10 km thick `waist'. Then the tensile stress produced in it by the submerging material of the plate increases by an order and it will inevitably tear apart too' (*idem,* p. 927).

Formation of marginal seas. According to all opinions expressed in plate tectonics, the subduction zones corresponding to the sites of plate collision are compression zones. This definition can be found in all works expounding plate movements. However, over the entire western and northern margins of the Pacific, the prevailing part of the area over seismofocal zones and behind them is occupied with marginal seas, recognized by most researchers, including all protagonists of plate tectonics, to be not only extension structures but spreading ones, i.e. those which are most active manifestations of extension. When the Sea of Japan, the Philippine Sea, the Barents Sea underwent their growth in the Cenozoic (and, for the most part, did so only in the latter half of this period, i.e. during several tens of millions of years), the island arcs in front of these seas moved away from the continent by 700—2800 km. The least removal distance is in the Sea of Japan whose age is the youngest, Neogene-Quaternary. Thus, the opening rate of the marginal seas can be estimated at 2—5 cm a year, i.e. of the same order of value as the ocean spreading and, corresponding to the propositions of the plate hypothesis, the same as the underthrusting in the subduction zones. So a question arises of why the rear of the seismofocal zone develops an extension of such a rate instead of compression due to subduction.

There are two attempts at explaining this discrepancy. According to the one suggested by Isacks and Molnar (1971) and supported by Le Pichon *et al.* (1973), extension of marginal seas is due to the fact that the forward part of the plate submerging into the mantle by gravity breaks away at a depth of several hundreds of kilometres to continue submerging at an elevated speed and thus creates an extension zone behind it. A supposition that the basalt plate submerges into a peridotitic mantle by gravity as the depth increases and, consequently, the mantle density grows (while the temperature of the submerging plate rises) is not plausible. Likewise, the figures that follow from this construction are also implausible: if, at the surface, a plate submerged at a rate of 9 cm a year (as was suggested, for example, by Le Pichon, 1968, for the Mariana Trench), at depth, it goes on moving inclined at 45° (which also defies explanation since it should fall down vertically by gravity), the rate of its submergence must increase up to $9 + 5 \times 1.5 = 16$—17 cm a year to cause a horizontal extension of 5 cm a year. If the direction of submersion is steeper, the new rate should be greater. But again, the extension of the margin sea crust proceeds not in the direction expected: not from the island arc towards the continent but from the continent to the island arc.

The other version suggests that the underthrusting of the plates generates a subsidiary flow of the mantle substance over it, and the direction of this flow near the surface reverses to move opposite to the plate and cause the opening of the marginal seas. Another supposition is that the submerging plate creates a mantle diapir over it which, while rising towards the surface, leads to a spreading extension of the marginal sea.

All versions prompt the conclusion that the island arc moves away from the continent at the rate of opening of the marginal basin to meet the underthrusting plate. In this process, the island arc (or the corresponding boundary, if it is believed that the arc is generated at some later stage) is to actively overthrust onto the underthrusting plate, making the bend line of this plate move oceanward, i.e. deforming the plate additionally and reducing the convection cell beneath it (if it exists). But the suggestion that a smaller mantle flows beneath the marginal seas, which are subordinated, forcing away the main flow when meeting it is not plausible energetically.

Shape of island arcs and trenches. According to the propositions of plate tectonics, the trajectory of plates corresponds to their rotation about certain poles on the Earth's surface (i.e. round the axis passing through these poles and the centre of the Earth). In so moving, all the points of the plate travel along the parallels drawn from the poles of rotation at a similar angular speed and the plate configuration does not change. Thus, the pattern of the plate consuming boundaries should replicate the pattern of the plate growth boundaries, i.e. of the spreading ridges. Such a correspondence is nowhere observed, and this alone is enough to testify to the discrepancy between the ideas of rigid plates and reality. It should be recognized that the plates are able to change their shapes and outlines considerably. However, even with such a retreat from the postulates of the plate hypothesis, there remains a great challenge to account for a characteristic feature of the island arcs, which is their horizontal bending towards the ocean. If this bending is considered to be due to different rates of underthrusting of plates in the middle of the island arcs and on their ends, then one has to explain the plate submergence at the greatest rate in the narrow zones inside the re-entrant angles between two arcs.

It is true, the horizontal bending of the island arcs can be given a natural explanation through a growth of the marginal sea admitted by plate tectonics, since the rate of this growth is maximum in the central parts of the marginal seas as follows from their shapes. However, if this were so, the influence of the growing structure of the marginal sea on the underthrusting plate would consist not only in the removal of the line of the oceanic plate bend but also in the change of this line with its consistent horizontal bending towards the ocean. The structure of the marginal seas thus turn out to be much more active than that of the underthrusting ocean plates.

The works on plate tectonics usually omit the issue of the convex shape of island arcs.

Double systems of seismofocal zones with similar direction of dip. On the sides of the Philippine sea one observes two seismofocal zones which are equally inclined to the west and towards the Asian continent. According to the subduction hypothesis, in both cases, the sea (oceanic) foundation should underthrust to the west, under the Mariana arc, and farther also to the west under the Philippine—Ryukyu arc. The ocean floor underthrusts under the Mariana arc in agreement with general subduction views , but the situation is much more complex inside the Philippine Sea. The eastern half of the foundation in this sea has to be considered as overthrusting on the Pacific floor. Underthrusting is possible for only the western half of the foundation, if the spreading line (according to the concepts of the supporters of the plate hypothesis) were drawn in the middle of the Philippine Sea (for example, over the Kyushu—Palau Ridge). In that case, a way out could be found (although a formal one) in constructing an asymmetric model of spreading: the western spreading wing of the Philippine Sea together with the Mariana island arc overthrusts over the Pacific arc, the eastern wing, that is, only a part of the sea foundation underthrusts under the Philippine Archipelago and the Ryuku Islands. However, this model is also impossible since, if the spreading zone expands symmetrically, then the western spreading wing should not be wider than the eastern one; but the latter remains wholly on the surface and the former has an underthrusted part submerged down to 1000 km beneath the Philippine Archipelago. It is rather hard to find an explanation for such a different width of foundation of the western and eastern parts of the Philippine Sea if they are products of spreading. To maintain the symmetry of the spreading growth, it is possible to shift the spreading axis close to the Philippine and Ryuku trenches. Such constructions would have little in common with the real structure of the Philippine Sea. Nor can one explain why seismofocal zones, trenches and island arcs in the western Philippine Sea, which were generated in such specific conditions, prove to be completely analogous to similar formations on the ocean margins where such uncommon plate underthrusting occurs.

Double systems of seismofocal zones with opposite direction of dip. Contradictions connected with the opposite dipping of closely disposed subparallel seismofocal zones are obvious enough and, thus, have been discussed repeatedly. What is meant primarily in these cases is the area of New Hebrides, Solomon Islands and Archipelago of Bismarck. This is a place where two of the six plates identified by Le Pichon — the Pacific and the India-Australian — move to meet each other. Instead of underthrusting one under the other, they underthrust both under the archipelagoes mentioned above, leaving a band 250 km to 1000 km wide between their trenches. Explanations of such a disposition of the plates which was not provided by the plate hypothesis can be found, for example, in the summary work of Le Pichon *et al.* (1973). These are, however, suppositions about changes of the direction of plate movement at depth, so as to correspond to the observed location of seismofocal zones. We will not dwell on these constructions at greater length because of their artificial nature. Among the mechanisms related to the surface structures, compression and suction are emphasized, since the plates come close to each other and both of them go down to depth. Compression accounts for the existence of islands while suction produces large marine basins between and near them. Distribution of compression is, however, not confirmed by a prevailingly gentle occurrence of Cenozoic deposits on the islands. As for suction: the large marine basins of the New Guinea Sea in the Bismarck Archipelago and North Fiji Basin have foundations with the usual geophysical properties of marginal seas, i.e. they have no signs of basins of special origin. The existence of these seas, as well as longitudinal graben-like basins in the group of Solomon Isles, indicate extension conditions, but not those of compression.

The situation in the Indonesian (Indonesian—Philippine) area is not less indicative. This area — the Sunda Isles and the Philippines — is disposed between two large trenches — the Sunda and Philippine Trenches — which are accompanied by very deep seismofocal zones inclined to meet each other (the Sunda zone is the world's deepest known zone with earthquake foci of up to 720 km), in contradistinction to the rather shallow seismofocal zones in the area of New Britain—New Hebrides. One should expect that where two areas of compression in counter directions come close together, geological compression forms would be the most evident. Instead, in the Indonesian area, we can observe a series of quite deep seas of marginal type — the South China Sea, Sulu Sea, Sulawesi Sea, Banda Sea, and smaller basins — together with obvious signs of widely destroyed continental crust. In the south-eastern Indonesia area, where the two above-mentioned seismofocal zones come so close together that they meet each other at depth and, consequently, where superimposition of counter compression zones would be inevitable according to the subduction scheme, the larger part of the surface is occupied by marginal type basins attesting to extension: these are the Banda Sea, Ceram Sea, Molucca Sea, Sulawesi Sea, Flores Sea, and the Makassar Strait.

Evidence of the occurrence of Cenozoic and Upper Cretaceous deposits on the Indonesian islands equally attests to conditions of no compression. The ideas of wide distribution of an intense Alpine folding in the Indonesia—Philippine Archipelago, which is typical of the works of geophysicists, is often exaggerated. The summarized information on the tectonic structure of the Indonesian Archipelago (for instance, from van Bemmelen, 1949) always leads to an opposite conclusion. Not only in the inner part of this Archipelago on Kalimantan, but also on the marginal islands of the Sunda arc, on Sumatra and on a considerable part of the Java, folded deformations of Paleogene-Neogene deposits are weak; indications of angular nonconformities inside the Cenozoic deposits, sometimes referred to as evidence for the phases of the Alpine folding, in fact largely relate to gently occurring sections that contain no large angular nonconformities. As all the researchers point out, an intense Alpine-type folding is only typical of narrow zones in the margins of the region as the Nias—Mentawai Islands in the west, and the Timor—Tanimbar Islands in the East. The latter is opposite the northern shelf of the Australian continent, and fold dislocations distributed in it should be associated with the consequences of northward movement of the Australian platform. It is not possible to characterize the development of the Indonesian megastructure as a whole by these dislocations. On the Philippine Islands, the Cenozoic deposits are almost everywhere only mildly disturbed.

The situation in the Indonesian Archipelago as well as in the other areas described above obviously contradicts the subduction hypothesis.

Generation of outer oceanic swells near trenches. Some researchers point to the existence of outer oceanic swells, considering them to be large-scale anticline folds — the result of an elastic bend of plates where they were underthrust. However, it would be reasonable to expect the appearance of a swell right before the underthrusting line, i.e. the trench axis, rather than at some distance from it. The position of oceanic swells is parallel to the trench axis, but they are disposed at a distance of 150-200 km oceanward.

Origin of heat flow. The anomalously high heat release over seismofocal zones is a challenge that seems hardly feasible for the subduction hypothesis to overcome. It would be natural to expect that submergence of a cold plate into the hot mantle might result in the cooling of the mantle and reduced heat flow above. Instead, seismofocal zones display a heat flow heightened two or more times, and magmatic activity that testifies to great contributions of heat energy from depth. An explanation for this phenomenon is attempted by postulating the heating-up of mantle due to friction on the boundaries of the submerging plate. This gives rise to one of the basic contradictions of the subduction hypothesis: the plate retains its form and capacity to go down into the mantle because it remains at relatively low temperatures— but at the same time it is to generate enough heat as to be able to heat vast volumes of mantle.

A number of works of the adherents of new global tectonics are devoted to computations of heat effects in the process of plate submergence. In most cases, the computations are grounded on the determination of the amounts of heat generated by friction according to the available heat flow and rather arbitrary physical parameters (Le Pichon *et al.*, 1973). However, attempts to compute a heat effect from friction based on even the most favourable assumptions lead only to uncertain conclusions (Toksöz *et al.*, 1971).

A characteristic calculation, for example, is given by L. Sydora *et al.* (1980) who determined the amount of heat necessary to melt 1% basalt out of the lithosphere volume above the level of 150 km. According to the conclusion reached, 1 m^3 of a friction-heated layer at the plate surface with a layer thickness of 5-10 km would liberate 120-320 MW of heat. Thus, in a matter of seconds, this cubic metre must liberate, as a result of friction, an amount of heat that is enough to melt it. The plate travels at a rate of several centimetres a year, i.e. a displacement per second of only some thousandths of a micron for a thickness of 5-10 km. Such a mechanism of heat generation seems unrealistic. However, according to L. Sydora *et al.* (1980), the slowing of plate velocity does not hamper the melting of basalts but facilitates it, as the calculated plate-liberated heat is integrated into the mass.

Ye. V. Artyushkov (1980) showed that if all the potential energy of a volume of cooled ultrabasic rocks, which sinks in the hot mantle down to a depth of 700 km, is turned into heat energy, disregarding its heating up and corresponding loss of density (i.e. obviously increasing the result), the rise in temperature of this volume will reach a mere 500° C. Naturally, the heat that is liberated due to friction when a body submerges by gravity (if this is possible) makes only a part of this amount.

Occurrence of sediments in trenches . The authors of the early works who aimed at reconciling plate tectonics with geological evidence, supposed that the trenches have accumulated crumpled deposits (e.g. Le Pichon *et al,* 1973). Such a conclusion is natural since the sedimentary layer on the oceanic floor is poorly lithified and is therefore mechanically weak compared with its underlying ocean crust and the foundation of the continental slope. Underthrusting of the oceanic foundation takes place under conditions of resistance offered by a motionless continental slope; in this process, deformations are inevitable (in particular, new global tectonics considers the pressure exerted by subduction to be a cause for the origin of such folding systems as the Alpine). In addition to poor strength and high plasticity of a sedimentary layer in the ocean, it also has a free surface and undoubtedly should, first of all, undergo deformation during underthrusting. It was supposed that those deformations consisted in scraping off sediments from the underthrusting oceanic foundation, to be followed by their piling up and intense crumpling.

In 1973, in the Report on deep-sea drilling carried out during the 13th Leg of *Glomar Challenger,* its authors said: `Drilling along the Hellenic Trench margins was anticipated to reveal deformation pattern of the carpet as it is scraped up where one lithospheric plate plunges under another...' (*Initial Reports...*, 1973a, p.5). The holes drilled near the trench axis revealed only non-dislocated deposits to a depth of 480 m without reaching older sediments or the basic foundation. The section of the Quaternary deposits at the foot of the continental (island arc) slope correlated with that of the axial part of the trench with accuracy to centimetre-thick layers (*idem,* p. 275). The authors pointed out further: `Le Pichon has suggested that the African Plate, of which the Mediterranean Ridge and outer trench wall are a part, has been underthrusting the European plate (Aegean Plate of McKenzie) at some 2.6 cm/year averaged over part of the Late Cenozoic... What can we say about this from the drilling results? First, the age of the stratified sediment fill in the present Hellenic Trench (> 1.8 my) is far in excess of that which would accumulate on a transient piece of Ionian Basin floor some 6 km in width (present trench width). A consumption rate of 2.6 cm/year would engulf

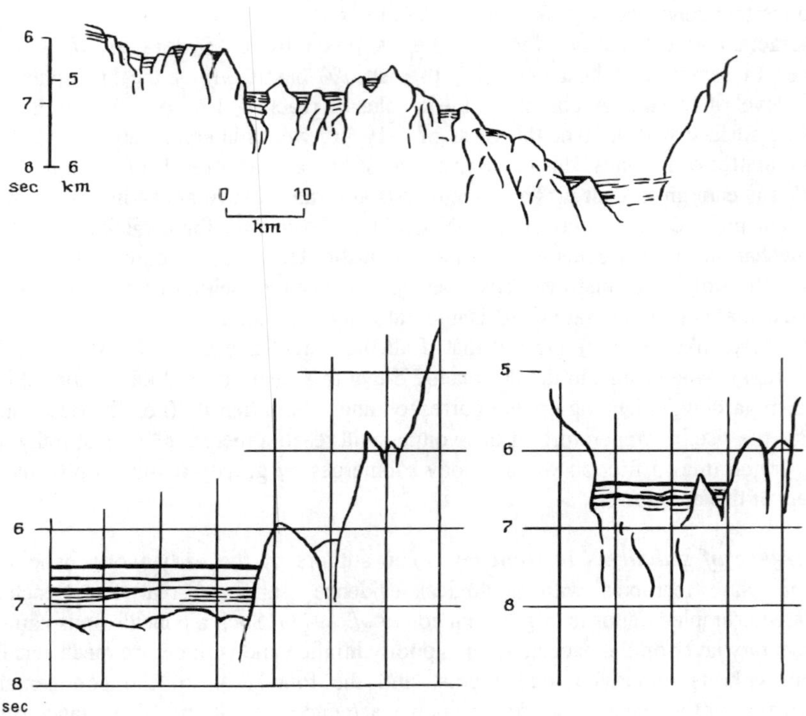

Fig. 23. Comparison of sediment occurrence on the slope of Mid-Atlantic Ridge and in Peruvian Trench, according to seismic profiling: A—axial and lower of outer slope parts of the Peruvian Trench (after R. Prince and L. Kulm, 1975); B—western slope of the Mid-Atlantic Ridge near its foot (after Vogt *et al.*, 1972).

the present flat trench floor in only 230,000 years...' So, the time of accumulation of the Quaternary cover estimated at Sites 127 and 128 of this Leg means that such an underthrusting of the sediment cover should have occurred eight times' (*idem*). In 1973, the discovery of non-dislocated sediments in the trench came as a surprise, but in the years that followed thefact of horizontal occurrence of sediments in the trenches became commonplace thanks to other boreholes and, mainly, thanks to a widely practiced seismoacoustic profiling.

It is worthy of note that the adherents of the plate hypothesis seek a proof for the oceanic floor subduction in trenches referring to signs of mild non-conformities among the subhorizontal sediments there. For example, R. Prince and L. Kulm (1975), who explored the Peruvian—Chilean Trench, (Fig. 16) even for the most ancient sediments within it, can point out only that they are gently folded (see also Fig. 23). In reality, such non-conformities, especially in the turbidite series which are distributed in the trenches, testify to no more than intraformational erosion. Furthermore, the data of Prince and Kulm are significant, as these scientists obtained evidence, according to their calculations, for huge vertical movements at the floor of the trench amounting to 14-22 cm a year.

Geologically, an absence of deformations of the sediments in the trenches is an adequate argument to admit an absence of subduction there. The author of this book believes that the stubborn adherence to the subduction idea in spite of this obvious circumstance is due to the fact that this idea is elaborated mainly by geophysicists, who do not perceive its proper importance, while geologists are not familiar enough with the geophysical research in the trenches. For comparison, Fig. 23 shows seismoacoustic profiles obtained from the Peruvian Trench and the slope of the Mid-Atlantic Ridge. In both cases the sedimentary cover as a whole, and in all its constituent parts, is equally horizontal and has only very local dislocations near the lines of displacements of the bed-rock foundation. It is impossible to see signs of tension in one case and signs of compression in others: the tectonic conditions in both cases seem to be quite similar.

It is symptomatic that there are no thick sediments in the trenches with the continental slopes rising 10-12 km over them, with a portion of the slopes outcropping onto the land surface and undergoing intense erosion. The sedimentation rate in such conditions should be considerable, which is confirmed through its direct determination: for example, in the Kuril—Kamchatka Trench, this rate found from water-contained suspension should be 20-30 cm per 1000 years (Lisitsyn, 1974). The same rate is typical of sedimentation over great areas in the Sea of Okhotsk and the Bering Sea where several kilometres of sediments accumulated during the late Cenozoic (4 km of sediments for 16 my, or 1 km for 4 my). There is no doubt that a number of the ocean trenches have existed for much longer but they have no sediments of comparable thickness. The great sedimentation rate in the trenches is also confirmed by the presence of buried trenches near the Atlantic and Pacific coasts of North America with a thickness of many kilometres of sediments which filled them up (Emery, 1965; Scholl, 1974). The existence of buried trenches testifies to the fact that after movements in the seismofocal zones there comes their rapid filling up with sediments. The underthrust conditions do not present any obstacles for such filling up; on the contrary, additional sediments are delivered from the open ocean to the trench. If about 1000 km of the ocean floor is underthrust, then the volume of delivered sedimentary material is large enough to fill up a trench one-and-a-half or two times; but the plate tectonics concept, as shown below, allows sometimes for an underthrusting of ocean floor of over 10 000 km or even more.

The problem of occurrence of sediments in trenches is often passed over by the subduction theory adherents, or alternatively they allow for accidental circumstances: overlapping of crumpled sediments with non-crumpled ones, retardation of the subduction process, etc. A single attempt at explanation of the principle for horizontal occurrence of sediments in trenches which the author knows is that by O. G. Sorotkin and L.I. Lobkovskiy (1976). These authors believe that `the sediments which found their way into a gap between the plates are under the effect of the mutually opposite-directed impacts: the motion of the underthrust plate carries them under the edge of the opposite plate while an excess weight of this edge squeezes the sediments out of the underthrust zone to the outside. It is from the result of these mutually opposite forces that the direction of the process development depends; if the force of drawing of sediments into the gap is prevailing then they would be fully consumed in the underthrust zone without any `scraping off'; otherwise the excess sediments will be squeezed out of the underthrust zone and crumpled in front of the edge of the upper plate.' ( *idem,* p. 5). As one can see, this is a strange way of posing the problem: it is assumed that the sediments are already in the `gap' between the plates while the actual question is what caused them to get there when they had previously been on the surface (and incidentally, a question also of why the `gap' itself formed between the plates). For no obvious reason, the crumpling of the sediments is attributed not to their collision with the foot of a motionless continental slope but to their squeezing out of the `gap'. With the help of the equation of Navier—Stokes, which governs the flow of incompressible liquid, and using triple integrals the authors arrive at the conclusion that if the thickness of the sediments is less than 400-500 m, then, while being in the gap, they will flow downwards following the underthrust plate, and if the thickness is greater, they would flow out backwards into the trench and be crumpled. It is unlikely that this model has anything in common with reality.

OCEAN-MARGIN ACCRETION AND SOME RELATED QUESTIONS

During its 28-year history, the concept of plate tectonics has been carried along by a succession of overenthusiasms that followed one another. All of them had, at heart, the purpose of making the plate tectonics concept more rigorous and concrete. Thus, during the early years after the creation of plate tectonics, it was popular to determine the position on the Earth's surface of the rotation poles of all the existing pairs of the plates, with numerous papers dealing with this subject. However, these determinations were a formal geometric description of a relative displacement of plates in terms of their rotation about a certain point on the globe surface. They contributed nothing to a concrete comprehension of the phenomena developing on the Earth and now they have virtually passed into oblivion. Another topic under consideration from the moment plate tectonics came into being was that of a triple junction of plates. It stems, in its essence, from the fact that in the case of the junction of three spreading ridges of different strike (there are several such points on the Earth's surface), one should explain the simultaneous displacement from that point of the three plates in three different directions; this is more difficult to do than to explain (on a single plane) the displacement in opposite directions of two plates relative to a single spreading line which divides them. Essentially, the triple junction of the spreading ridges inevitably leads to the conclusion of an extension of the ridges along their axes, since the three vectors of the forces radiating from the junction point and affecting the oceanic foundation form pairs, with the resultant forces directed along the axes of the ridges from the junction point. However, this conclusion was not usually arrived at.

In the 1970s to early 1980s, most attention was paid to `hot spots' which were supposedly formed by `mantle plumes'. It was expected that the hot spots, travelling over the plate surface due to the moment of the latter, leave on the Earth's surface a trace in the form of extended lines of magmatic formations and these traces allow us to make a judgment about the direction and rate of the plate displacements. However, these hopes were not fully realized. At the moment, the supposition persists that there is such a trace in the northwestern part of the Pacific in the form of the Hawaiian Islands and their continuation – the submeridional submarine Emperor Ridge (critical treatment of the supposition can be found in the following chapters; see also Scalera, 1991, 1995). There is less certainty when the ridges adjacent to the Tristan da Cunha Islands in the Atlantic Ocean are assigned to such a trace. But attempts to find material for similar constructions on the continents failed. On the contrary, it became clear that those objects most suitable to be treated as hot spots (for example, the area of the Yellowstone National Park in the USA) retain a fixed position within the `plates' for a long time.

The most recent work which specifies the plate tectonics model was a study of a process related to active ocean margins and which was termed `accretion'. It should be noted that originally this term was used in plate tectonics in another sense: it implied a growth of plates in spreading ridges (for example, Le Pichon *et al.,* 1973). The term `accretion' is currently used to mean separation from the underthrusting plates of blocks and slabs which do not move to under the opposing plate but overthrust onto its edge. So, the accretion process is essentially similar to obduction, but it is imagined as a process capable of forming a more extensive structure. Such an accretion follows the subduction and supplements its previously constructed model. Development of the accretion concept stems from the works by W. Dickinson (1974), D. Seely (1977), Seely *et al.*,(1974), D. Karig (1977), and other authors.

As a result of the heaped accretional slabs in front of island arcs and the active continental margins, an accretional wedge is formed consisting of blocks of oceanic sediments and oceanic foundation. According to the views which have recently become common, these blocks can make up not only the feet of the continents and island arcs but also their slopes as a whole. Further development of accretion ideas has brought about a new concept on the formation of the island arcs themselves and vast marginal parts of the continents adjacent to the active continental slopes. Some authors consider that accretion is the basic process which leads to transformation of the oceanic crust into the continental one (e.g. Parfenov, 1983). In recent years, accretion has come to be considered as the cause of the magmatic, ore-forming and other processes taking place near the active oceanic margins. So, accretion zones have now become a place of concentration of all the most important processes occurring on the Earth.

An extreme embodiment of the accretion idea is the suggestion of a combination of heterogeneous terrains within the complexly built margin areas, especially around the Pacific as a basis for comprehension of their general structure. This leads to doing away, to a considerable extent, with the question of regularities in the development of such provinces. Chapter IV shows, however, the possibility of not such chaotic, but historical comprehension of the development of such very tectonically complex areas as the regions of northeastern Russia.

Substantial criticism of the accretion concept was made some time ago by D. Scholl and co-authors (1977). Their arguments are based, first of all, on the obvious discrepancy in thickness and composition of deposits of the oceanic basins and of the continental and island arcs slopes that are adjacent to the trenches. The sediments in the open ocean are of small thickness (several hundreds of metres) and are represented mainly by the oceanic, essentially biogenic oozes and deep-sea clays; the sediments on the continental slopes are represented by turbidites and other rather coarse material of several kilometres thickness. It is untenable to suppose that the latter are sediments of the oceanic basins moved out of the ocean. Where, according to the interpretation of seismoacoustic profiles, highly deformed formations exist under horizontally bedded or weakly deformed sediments of the continental slopes, (for example, on the island arc slope of the Sunda Trench), they embrace vertical sections down to 10-20 km, as some authors point out. No doubt, the sedimentary masses forming those volumes, if they exist, are much thicker and of different composition as compared with a 400-metre section of the oceanic oozes and clays occurring on the oceanic side of that Sunda Trench. There are also no grounds to assert that such sediments are common in the folded mountainous ranges adjacent to the active margins; on the contrary, they consist mainly of rather coarse sediments whose features are close to the turbidite accumulations of the continental slopes. In many cases, (for example, in the coastal ranges of Chile, South Peru and Mexico) such sediments are Paleozoic, overlying stratigraphically more ancient complexes of continental-type rocks and are, no doubt, authochtonous.

The arguments of Scholl and co-authors are convincing. One cannot agree only with the statement on exclusive conditions when the axial parts of the trenches are filled with thick sedimentary prisms of turbidite character. Relying on the conclusions obtained earlier by Scholl in investigating (according to seismic evidence) the currently buried Washington-Oregon Trench (Scholl, 1974), they consider that such a burial was possible only in conditions of the Quaternary glaciation when the ocean level dropped abruptly and erosion of the coastal land became more intense. However, it is more probable that thick sedimentary prisms in the trenches appear regularly when the tectonic activity that creates the trenches ceases, irrespective of climatic changes. As the filling of the trench which stopped its active development occurs from the side of the continental slope, it is natural that the filling material corresponds to the sedimentary cover of the latter. The material which forms the cover of the present-day continental slopes and fills the buried trenches, on the one hand, and the oceanic deposits distributed beyond the outer slope of the trenches, on the other hand, just are the sediment complexes which should be opposed to one another.

The objections by Scholl and his co-authors should be supplemented with another one that the fold dislocations of the sedimentary cover on the continental slopes of the present-day trenches are not, in the main, complex or intense; they are as a rule discontinuous and normally appear near the faults dividing the basement blocks (see Figs 8, 9, 10, 11, 12 and similar seismoacoustic sections). These dislocations do not testify to large-scale lateral overthrusting of masses and general compressive conditions. As to deformed masses observed in the deeper parts of some sections underlying the sedimentary cover, which are often only supposedly interpreted as folded sedimentary rocks, even if they are actually folded, they are always separated from currently exposed surface sedimentary cover with a large angular nonconformity. Such nonconformity of regional character contradicts the idea of accretion, a phenomenon which should be continuous enough, and it marks, most probably, the base of the sediments overlying (after a long stratigraphical hiatus) an ancient (Paleozoic or Pre-Cambrian) folded foundation. The fact that such foundation outcrops the recent continental slopes is established by dredging.

The concept of accretion, however, is based on more factual material than most other divisions of plate tectonics (with the exception of the spreading in mid-ocean ridges). The seismic sections do show that the foundation of the continental and island-arc slopes consist of very complex formations; the same is also testified to by geological observations which are elaborated in more detail in Chapter IV of this book. On the whole, the works dealing with accretion and related themes (for instance, with the relationship between terrains in the northwestern and northeastern surroundings of the Pacific) expound specific geological and geophysical evidence in much greater detail than is common in plate tectonics publications and contain a highly professional geological analysis (for example, in the works by L. M. Parfenov, S. D. Sokolov, M. Churkin, and others). But, therefore, it is all the more necessary to recollect that an idea, similar to accretion, involving the heaping of scraped, compacted and deformed material where underthrusting of plates occurred was expressed at the dawn of plate tectonics with respect to sediments of the open ocean being transported in the trenches. Seismoacoustic profiling disproves this suggestion, but now the same idea has been transferred over the non-deformed sediments of the axis trench zone within the continental and island arc slopes where deformed sediments are easier to find. No explanation for the absence of deformations in the trenches has been proposed, and the significance of the non-deformed zone is ignored. This is illogical if one imagines the plates underthrusting from the side of the ocean, and this is the main objection against the accretion of the simultaneous with seismofocal zones oceanic material beyond the trenches.

Of great interest is the distribution of ultrabasic rocks and gabbroids as well as lawsonite-glaucophane and similar amphibole (actinolite, tremolite, hastingsite) slates which are revealed by dredging and deep-sea drilling near the foot of the continental and island arc slopes of many trenches (Kuril—Kamchatka, Mariana, Puerto Rico, Central-American). Special attention is often paid to `blue' glaucophane schists that testify to anomalously high pressures at shallow depths.

`A determining sign of an accretion wedge', L.M. Parfenov, for instance, writes (1983, p.5), 'which distinguishes it from other formations wherein ophiolites participate is the presence of lawsonite-glaucophane schists in the composition of tectonic flakes as signs of superhigh pressure typical of the Benioff zones'; similar opinions were expressed by many other authors. A question arises, however, of how the lawsonite-glaucophane schists which originally were in a comparatively deep part of the Benioff zone are brought up to the Earth's surface. The depth of their origin should be more than several kilometres; their emergence on the surface calls for material elevation from depth rather than of carrying it away into the depth. This is one more weakness of the plate tectonics concept: an idea of trapping and carrying away into the depths of near-surface masses obviously contradicts the observed appearance on the surface of depth genesis rocks, and just everywhere in the areas of the outcrops onto the Earth's surface of the main surfaces of seismofocal zones. This is not only true of blue schists but of the entire complex of basic and ultrabasic rocks which most often enter into the composition of the ultrabasic melange. Thus, the wide distribution, in itself, in the areas of recent and extinct seismofocal zones of ultrabasites and ultrabasite-containing tectonic mixtures (melange) where such undoubtedly depth-originated rocks as eclogites are rather common, testifies to the uplift of deep-seated material, including the mantle one, in spite of the accretion idea of plate tectonics. It is much more plausible to think that the complex composition of continental and island arc slopes at the active ocean margins may be due to depth accretion, which might indeed exist, but connected with a direction of main tectonic movements that is the very opposite of subduction.

STRUCTURE OF THE SUNDA TRENCH—THE HIMALAYAS

In many cases, along with other arguments in favour of subduction, an argument is offered for an almost uninterrupted border which is presented in the south by the Sunda Trench and the line of the Nicobar and Andaman Islands, and in the north by an overthrust zone of the southern foot of the Himalayas. A similar continuity of analogous borders can be observed in some other regions (the Hellenic Trench — the northern front of the Arabian block), but in the Sunda Trench and the Himalayas it is most pronounced. The suggestion is made that, since the tectonic structures observed at the foot of the Himalayas indicate underthrusting of the Indian Platform under the Himalayas (or an overthrust of the Himalayas), an underthrusting in the same direction should also take place along the Sunda Trench.

However, a difference in character of the border in front of the Himalayas and the Sunda Trench is quite obvious. In the former case, a part of the ancient continent has been underthrust while in the latter case it is supposed that the ocean floor does so: in the former case, the border of underthrusting is disposed at a shallow depth beneath a thick sedimentary cover; in the latter case it is on the floor of the deep trench with thin sediments. In the area of the Indian Ocean adjacent to this trench, deep-sea drilling has revealed sediments of the Early Cretaceous (Albian) age (and in the area adjacent to the Australian shelf they are of Late Jurassic age). This means that this part of the Indian Ocean had already existed when the Indian platform was far to the south of the present-day line of the base of the Himalayas and was just starting its travel northwards at a rate of up to 10 cm per year as follows from the paleomagnetic measurements (McElhinny, 1968; and other publications). Between the Indian Platform, which previously was a part of Gondwanaland, and the Asian continent, which included Indochina and Northern Tibet, there existed an open oceanic space — the Tethys; a part of the Indian Ocean adjacent to the Sunda Archipelago was, thus, a part of the Tethys and today remains its relic (like the Black Sea and the South-Caspian Basin). The Indian subcontinent, thus, invaded the Tethys destroying, folding and pressing into the depths the ocean foundation along the front of its movement. The latter process of pressing and submerging the foundation can be referred to as subduction, but in another, narrower, sense than what this term implies in plate tectonics. Precisely such a concept on the movement of the Indian subcontinent over the Mesozoic Tethys due to the effect of the opening to the south of the Mid-Indian spreading ridge was offered by R. Dietz and J. Holden (1971). The travel of the Indian Platform through the Mesozoic-Early Cenozoic Tethys is a quite opposite and later process compared with the formation of the Tethys basement. The character of the latter one can judge from the structure of the preserved part of Tethys, i.e. the northeastern part of the Indian Ocean opposite the Sunda Trench, but not from the processes which took place in front of the movement of the Indian Platform, between it and the Himalayan area, although this front is now near the line corresponding to the northern border of the Mesozoic Tethys. The history of the development of that part of the Indian Ocean can already be expounded from the data of the deep-sea drilling of the *Glomar Challenger*. The data obtained on the age of the sedimentary cover testify to the growth of the oceanic floor from the Sunda Trench at least from the Albian and, respectively, to no underthrusting in this trench (see Chapter V).

Chapter III. Eduction

MAIN PREMISES FOR THE EDUCTION CONCEPT

If a geosyncline or a marginal sea opens by moving apart of the wings of a large fault zone, a question arises of how the space between the separated parts of the Earth's crust is filled up. The foundation which is formed both in geosynclines and marginal seas is in most cases oceanic, i.e. it has no geophysical granite layer and is made up near the surface of basic material alternating with some amounts of ultrabasic rocks. Formation of oceanic basement in the oceans, at least in a large part of them, has found its explanation in spreading. It is common to extend the same mechanism to marginal seas too in order to explain formation of their basement. However, in most of the marginal seas, there are no middle ridges and the seas themselves are of a clear asymmetric pattern: on one side of them there is an island arc with a seismofocal surface beneath it dipping under the marginal sea; on the other side there is a continent or, as is the case of the Philippine Sea, another island arc with one more seismofocal zone which is again sloping towards the continent. In such a sea, arrival at the Earth's surface of a deep-seated material, necessary to build up an expanding oceanic-type foundation of the marginal sea surface, should also proceed in an asymmetric way. The mantle and the oceanic basaltic layer disposed on the ocean side adjoins most closely the marginal sea but they are separated by a deep seismofocal zone, so that the direct displacement of the substance of the oceanic wing of the zone to the area of the marginal sea is hardly possible. Such a displacement would make the existence of the very marginal seismofocal zone impossible. The mantle which is on the continental side of the marginal sea alone can be immediately connected with its foundation; should there exist another seismofocal zone, as is the case in the Philippine Sea (such seismofocal zones could also exist in some other marginal seas, as is suggested by buried trenches), it would be sloping from the marginal sea which means that the deep mantle material necessary to form its foundation could arrive right from the hanging wing of that second seismofocal zone. If we imagine the beginning of the origin of a marginal sea when the initial break-up of the Earth's crust took place and the split-off continental edge (later an island arc) started moving away and we exclude (as argued above) the possibility of spreading, then the only way of filling up a space between the forced-apart crustal blocks would prove to be the drawing of deep material from under the continent's edge.

This solution has already been pointed out by some authors. For example, the Australian geologists G. Packham and D. Falvey (1971), while adhering to the plate tectonic hypothesis, assume that in the marginal seas, deep material arrives mainly from under the continental edge (Fig 24). To explain the origin of the marginal seas' foundation, the authors advance a concept of an `asymmetric spreading' which finally leads to a conclusion that the mantle material elevates from under the continent. They link this process and the very formation of marginal seas in the Western Pacific to a restructuring of the Earth's surface in the Eocene — Early Oligocene, when the Indian subcontinent became connected with Asia: this resulted in the moving of the mantle material to beneath the Asian continent and its outflow from under its eastern margin. Outflow of the mantle material from under the continental edge to the marginal sea involves the concept of eduction, but Packham and Falvey do not attribute it to the existence of seismofocal zones.

Another example of an eduction approach is offered by a scheme of development of the island arcs and oceanic trenches suggested by M. Hoshino (1983). He presents the island arc in the form of a crustal prism from which the mantle material flows out to either side (Fig. 25).

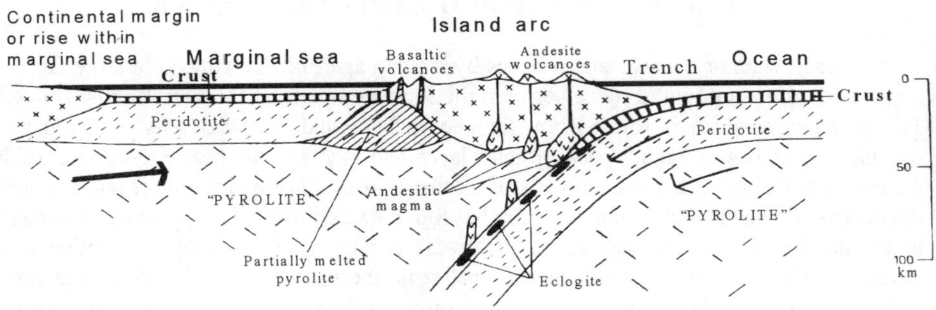

Fig. 24. Schematic cross-section of an actively developing marginal sea (after Packham and Falvey, 1971). Arrows — direction of movement of mantle matter; a supposed mechanism of formation of marginal sea basement results from active elevation of the mantle material from beneath the continent.

Where the mantle material outcrops onto the surface, there arise linear depressions in the sea floor, including a trench on the outer side of the arc which is conditioned by the highest rate of outflow of mantle material. An isostatic model for such a mechanism of trench formation is presented below. It should be noted that according to Hoshino's model, the outflow of most of the mantle material takes place on the outside of the island arc, i.e. along the main seismofocal surface. If this model is applied to the Philippine arc (Hoshino mostly used the data from the Japan Arc), then formation of the Philippine Sea basement took place by the outflowing of the mantle matter from under this arc. Hoshino's model lends itself well to island arcs of the Bismarck Archipelago—Solomon Islands which are on the two sides confined with the lines of trenches and seismofocal zones sloping towards each other. This model seems much more natural than are the constructions advanced by the supporters of the plate underthrusting hypothesis.

The outflowing of the mantle matter from under the continental edges is also assumed by the adherents of the hypothesis of an expanding Earth, since the continents do not take part to any great extent in expansion of the Earth's surface: the latter is due to expansion by the arrival of mantle matter to the surface of the oceanic space. So the mantle material which is beneath the continents should outflow towards the oceans (Koziar, 1980). The author (Chudinov, 1979, 1981, and other works) used this opinion to substantiate the eduction concept.

The mentioned viewpoints lack a connection between the notions of the outflow of the mantle matter from under the continental edges and the existence of the seismofocal zones. Identification of the phenomenon of outflow with the processes generating deep seismofocal zones is the crucial point in the view advanced by the author.

The first publication dealing with the eduction theme as a general phenomenon, but without use of this term, belongs to M. Keith (1972) who advanced a `reverse' model of plate tectonics according to which the same plates which were identified by Le Pichon, experience movement not in the direction towards the trenches but, on the contrary, are thrust out in the trenches and submerge in mid-oceanic ridges. Keith's work is fully devoted to the substantiation of such

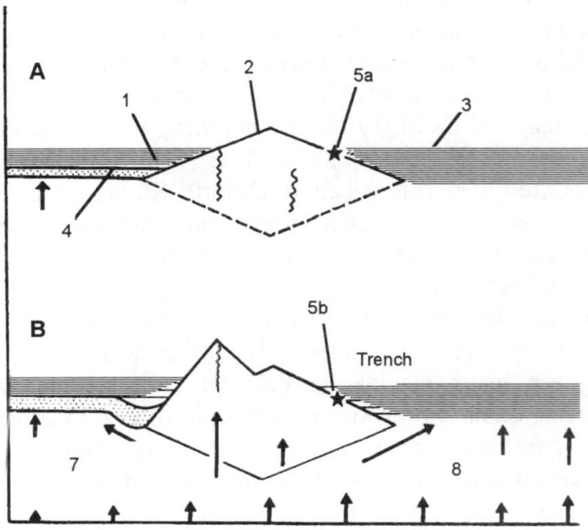

Fig. 25. Emergence of the trench near island arc, (after M. Hoshino, 1976,1983). A—first stage, single arc: 1—marginal sea, 2—island arc, 3—ocean, 4—sediments, 5a—terrace under formation; B—second stage, double arc: 5b—the same terrace deep under sea, 6-7 crust and mantle (the lithosphere): 6—of transient type, 7—of oceanic type. Arrows—directions of movement of mantle matter, wavy lines—paths of magma and fluid travel.

submerging. The suggestion of submergence of the oceanic crust in the middle ridges seems to contradict the facts, and Keith did not return to this model later on. Yu. M. Sheinmann (1973) cited Keith's scheme as an example of a construction showing the arbitrary character of the grounds for plate tectonics.

The term `eduction' was proposed by the author (Chudinov, 1979) and is derived from the Latin `educo', similarly to known terms `subduction' and `obduction'. The idea of such comprehension of geological processes connected with active boundaries of continents and oceans was conceived on the basis of the author's field explorations, mainly in the Russian Northeast, and analysis of other materials available.

After the author's publications (in the Abstracts of Papers of the XIV International Pacific Congress, Khabarovsk), the term `eduction' was used by the Canadian researchers J. Dixon and E. Farrar (1980) in the paper where the origin of the Franciscan melange in California was explained as a consequence of out-thrusting of mantle material from under the North American continent. The cause of the eduction was attributed by Dixon and Farrar to the existence beneath North America, according to wide-spread opinion, of a hidden extension of the Eastern Pacific mid-oceanic ridge which there continues its spreading. Accordingly, these authors consider eduction to be an exclusive phenomenon which is not connected regularly with the existence of seismofocal zones. In this paper, there is a reference to the authors' earlier abstracts where the term `eduction' was also used, but possibly in a different sense, as that work was devoted to Cenozoic history of the Colorado Plateau (Dixon and Farrrar, 1978). In 1981, the author of this book published another paper (Chudinov, 1981) about eduction, much larger than the previous one (received by editorship of *Geotektonika* in 1977). Two papers were also proposed by the author for the 29th International Geological Congress (Chudinov, 1992a, 1992b).

In recent years, an interest in eduction problems has been rekindled. At the beginning of the 1990s, several papers were published by different authors from different countries devoted to eduction, and others to `exhumation' and uplift of the depth material (mainly in active marginal zones), a phenomenon which is close to eduction (see Andersen, 1991, Avigad and Garfunkel, 1991, Fyfe *et al.*, 1991, Menier *et al.*, 1991, Barriga *et al.*, 1992, Little *et al.*, 1992, Martignole, 1992, Okay and Cedal, 1992). Among them, and devoted to eduction proper, are works by T. Andersen and W. Fyfe with co-authors (Fyfe *et al.*, 1991, Barriga *et al.*, 1992). Andersen considers the Caledonian eduction in North Norway in a sense close to mine (an outthrusting of deep material in the paleoseismofocal zone) and to Dixon and Farrar (as an exclusive phenomenon on the background of subduction). In the opinion of Fyfe and co-authors "mantle eduction is a second order process in plate tectonics" (Fyfe *et al.*, 1991, p. 419); however, in their interpretation, eduction is efflux of mantle gaseous and magmatic material rather than tectonic out-thrusting. The other works mentioned treat uplift and `exhumation' of deep mafic material in different parts of the world: Aegean region, Scandinavia, Australia, China, North America and some others in various geological times. The term `exhumation' does not seem to this author to be appropriate as it implies a previous submergence of risen material, which may be an unwarranted assumption..

CONSEQUENCES OF EDUCTION

According to the above discussion, eduction implies out-thrusting of the mantle material towards the Earth's surface from the continental edges and island arcs and occurs in marginal oceanic trenches. Eduction engenders deep-seated seismofocal zones. Supplementary to the spreading, it results in a growth of new oceanic crust from the side of the trenches in the parts of the oceans adjoining the continents and in the marginal seas.

It was noted above that the geological phenomena taking place in the areas where seismofocal zones have developed prove invariably to be the reverse of what should be logically expected if one proceeded from the subduction concept. It follows naturally, however, that these events are compatible with eduction. Before drawing a comparison between eduction concepts and actual geological and geophysical materials, we should explore a little further the logical consequences of the eduction process.

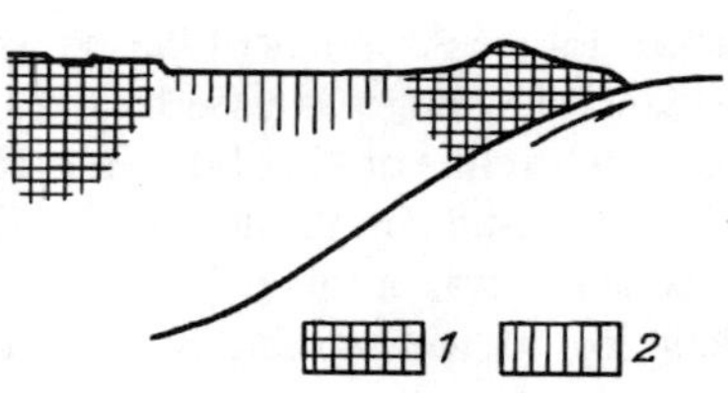

Fig. 26. Eduction model of the origin of an island arc and marginal sea (marginal geosyncline), by the author. 1—Continental lithosphere, 2—area of its rupture formed by tension.

Kinematics within above-seismofocal areas. Since the main movement of the Earth's matter in eduction consists in an outflowing of deep material from under immobile continental edges, these latter are subjected to an applied tension from below. In many cases this should

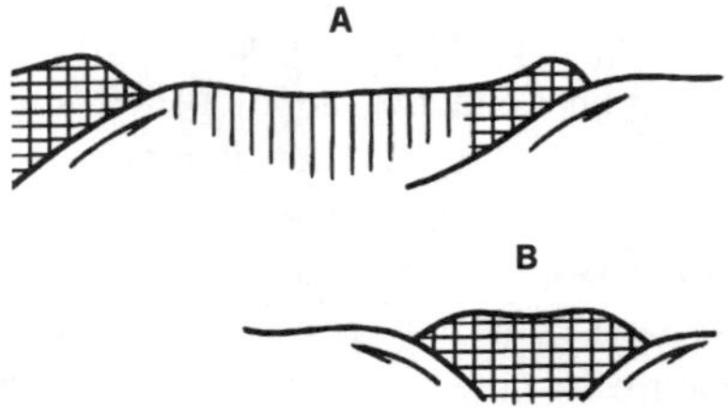

Fig. 27. Eduction origin of double systems of seismofocal zones. A — with similar dips (as in the Philippine Sea), B — with opposite dips (outflow of depth matter on either side of an island arc block). For symbols see Fig. 26.

result in the breaking off and separation of the marginal portions of the continental lithosphere from the main continental massif and their transportation towards the ocean. Behind such detached fragments of the continental lithosphere, areas of separation of the continental crust should be formed which should be filled up with material issuing from the depths. Many researchers have proposed such an explanation of the origin of island arcs and marginal seas behind them, but they have always been challenged with the causes of this phenomenon: eduction is the mechanism underlying it. Contrary to the subduction model, the effects of activity at depth upon tectonic displacements near the Earth's surface are direct and call for no additional complicated constructions (Fig. 26).

The existence of an extra seismofocal zone with its dip parallel to the main one, as well as the existence of a trench line on the inner side of a marginal sea, can also be accounted for eductionally without contradictions. The opening marginal sea needs its own mechanism for arrival of depth matter, which fills up the space created by thrust-apart crustal blocks and forms a new basement of the oceanic type. Eductionally, a seismofocal zone is just such a mechanism which is widely distributed and, thus, is energetically profitable. Therefore, the presence of a seismofocal zone at the inner side of a marginal sea (along with a trench and, occasionally, an extra island arc, as in the case of the Philippine Sea) is a natural consequence (Fig. 27, A). One could suggest that such seismofocal zones came into being at certain development stages in other marginal seas, which is evidenced by the seismic data on the existence of buried trenches, for example, on the inside borders of the Bering Sea (Schmidt, 1980). An inclination of the inside seismofocal zones of marginal seas towards the continent, i.e. in the direction from whence the depth material originates again follows naturally.

At the same time, the dip of the seismofocal zones of eductional origin is not preconditioned so rigidly as it is in the model of plate subduction. The elevating, educting masses create a system of various mantle flows of the one-sided direction from the depths to the Earth's surface; the tectonic surfaces that control them, accompanied by seismofocal zones along which the depth material outflows, depend for their position on a complicated and heterogeneous structure of the Earth's upper envelopes and can vary widely as a result. One can imagine an outflow of deep material not only from under the continent but also from under island arcs simultaneously on both sides, as is shown in Hoshino's model (Fig. 25). Such an outflow is possible because island arcs are detached pieces of the continental lithosphere and, consequently, preserve its properties to some extent. Similarly, one could explain the existence of counter-dipping seismofocal zones on both sides off the Solomon Islands and New Hebrides, which are amid oceanic space (Fig. 27, B). The same relationships between the dips of seismofocal zones can arise when the opposite sides of a large continent converge (i.e., when, correspondingly, part of it thins out in plan), if there were outflows of depth material from under its two sides simultaneously. This could account for counter-sloping seismofocal zones on both sides of the Indonesian Archipelago (among which, the Sunda Trench, as was mentioned, is a remnant of the marginal eduction zone of the Mesozoic Tethys). The fact that the extension structures prevail over the counter-dipping seismofocal zones — in the Indonesian Archipelago, the Bismarck Archipelago, New Hebrides-Fiji area — offers a confirmation for eduction motion in seismofocal zones.

One should also relate the shape of island arcs to kinetic manifestations of the eduction process: their most specific features — their convexity towards the ocean and their forming chains of festoons with entering from the ocean side angles where the arcs are closest to the continents — are not explainable using other geotectonic hypotheses. We will dwell on this problem once more below when we consider eduction-produced deformations, as well as the possibilities of large compression deformation during eduction in addition to that arising from tension.

It is fairly evident that the specific features of sedimentation in the marginal oceanic trenches, i.e. sporadic distribution of sediments, their poor thickness and weak deformations, are also highly compatible with the concept of eductional out-thrusting of the ocean floor from under the continental lithosphere; the sediments at the foot of the continental scarp do not have enough time to accumulate in large amounts, as they are perpetually carried away towards the ocean. Also, the necessary conditions are not present for their pressing and folding within the axial part and outer slope of trenches, quite similar to what is observed within the slopes of the mid-oceanic ridges. Local dislocations of sediments, if they occur, are due to the local, mainly vertical, block motions.

Emergence of high heat flow. The main challenge for the subduction model is how to account for the sources of heat flow arriving at those areas located above seismofocal zones: the solution to this problem within the eduction concept follows as a natural consequence from the model adopted. During eduction, the mantle material rises from depths down to 700 km (the deepest level being reached by seismofocal zones), where the initial temperature is 2000° C or more (e.g. Magnitsky, 1965; Moiseenko and Smyslov, 1986). When rising, this material is transported steadily to higher levels being kept at temperatures that are anomalously high for these levels because of slow cooling at such depths. Since eduction mantle currents involve vast volumes of mantle matter in the lying wing of the seismofocal zone, instead of a more defined slab in subduction, the possibility of delivery of the depth heat to the Earth's surface is very high. In essence, recognition of eduction leads to the quite new views on the amount of heat energy in tectonic, magmatic and ore-forming processes in above-seismofocal areas. If the eduction process is real, the consumption of heat energy in magmatic and ore-forming processes, as well as their temperatures, may prove to be much higher than assumed in available geological theories.

When discussing eduction problems, the author has sometimes come across an objection saying that the heat consequences of the eduction process contradict the fact that the lowest heat flow is to be found near the foot of the continental slope and in the axial zone of the trenches especially. In this connection let us remember, partly repeating what was said in Chapter I, that axes of trenches are a place where the mantle material of the lying wing of eduction zones appears for the first time on the Earth's surface, as follows from all geological, geophysical, geomorphological and other data. Before out-thrusting, this material had travelled under a volcanic belt, where losses of heat were already great; then, on its way to the trench floor, its rate of cooling increased progressively as it approached the Earth's surface. When out-thrust material appeared on the trench floor, its surface became subhorizontal, and consequently was bent, so the new eductional basement immediately underwent extension (which occupied also the closing part of the continental or island-arc slope). Thus, in the out-thrust material, a large number of open fissures must inevitably originate which are accessible to circulation of oceanic waters. As was shown above, this is a real phenomenon, which has been pointed out by researchers of heat flow on the ocean floor. The tensile conditions are confirmed by development of longitudinal grabens within the outer slopes of the trenches. According to Smirnov (1980), Smirnov and Sugrobov (1979-1980), and Smirnov *et al.* (1982), the low values in the average of the heat flow in the trenches are very likely to be associated precisely with the circulating sea water in the trench basement which has collapsed as a result of active seismotectonic processes, and also are a result of landslides and suspension flows. This fully conforms with the variable values found for heat flow, occasionally as high as 1.5-2 hfu and more (Chapter I), as well as with the negative values found in the Kuril-Kamchatka Trench, which were also mentioned above.

Water circulation in tectonically crushed rocks seems to be the main cause of great decrease in the apparent heat emission in the trench regions, as the boundary of the true heat flow descends to a depth out of reach for measurements. It is quite probable that the circulation leaves its evidence in the form of numerous fissures filled up with carbonates, zeolites and other low-temperature minerals, which are typical of the rocks within and near generated by eduction basement in the Russian North-East and other places.

INTERPRETATION OF SEISMOLOGICAL DATA

Seismological data are not only a source of basic information on seismofocal zones but they are sometimes used also as the main arguments in favour of the subduction concept. Geologists sometimes believe that seismological data unambiguously confirm subductional movements in seismofocal zones; however, this is not so. Any tectonic model dealing with active margins and seismofocal zones must, of course, be consistent with all seismological evidence.

In this section, we shall be concerned only with the grounds of seismological data used for geological interpretation. They have not changed basically from the 1950s and 1960s, when attempts at tectonic interpretation of seismofocal zones using seismological data started. To the present day, the same double-couple solution of earthquake mechanisms is most widespread, although the methods for their computation are often more complex. The history of these attempts is of appreciable interest.

Subjects and methods of investigations

In addition to the geographical location of earthquake foci, which was discussed above, seismological data are used to determine the direction of displacements in these foci (earthquake sources), and of stresses which cause the earthquake, usually using the supposition (supported by geological facts) that earthquakes are the consequences of displacement of material along a fault plane in their foci. These determinations make use of the fact that the first arrivals of longitudinal earthquake waves can be of compressional and extensional phases, and the areas of their distribution, when they are recorded by a number of stations, prove to be situated in a regular pattern; in some cases, after correction for curvature of the seismic waves, these areas can be divided by nearly perpendicular lines so that distribution of tensional and compressional first arrivals (or dilatations and compressions) can be interpreted as in quadrants (Fig. 28). The lines are supposed to represent the outcrops onto the Earth's surface of two mutually normal (or nearly normal) planes which pass over the earthquake focus, these are called nodal planes. One of the nodal planes is taken to coincide with a rupture at the

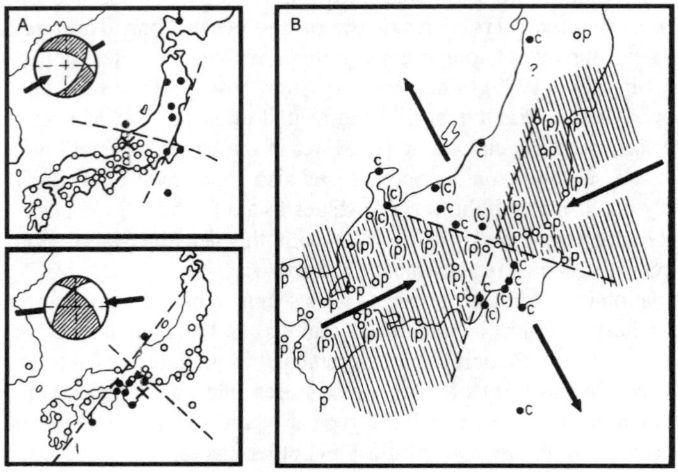

Fig. 28. Patterns of quadrant distribution of compressive and tensile first arrivals of longitudinal earthquake waves in Japan (after Honda and Masatsuka, from Richter, 1955). Compressive arrivals — black circles, tensile—white circles. **A**: in left-hand top corner the spatial positions of nodal planes and supposed orientation of compressive stresses are shown; **B**:—three degrees of determination confidence (no brackets, in brackets, with question mark) are shown; p—tension, c—compression. The method of estimation of the position of nodal planes remains essentially the same up to recent time.

earthquake focus. The position of areas of the initial compression arrivals of seismic waves indicate the direction of displacement in the earthquake focus, while the position of areas of initial tension arrivals point to the opposite side. So one of the nodal planes is taken as the fault plane. The question of which of two nodal planes this should be is mostly resolved from only circumstantial considerations (excluding some methods which are, in addition, records of the shear waves — these are more complex and less widely adopted). One does not always observe the quadrant pattern of seismic tensions and compressions; for example, according to Hodgson and Stevens (1964), this pattern can be conclusively proved for only one-fourth of the earthquakes investigated by various authors.

The summarizing investigations of the seismological data connected with earthquakes in seismofocal zones were, for a long time, aimed at identifying the prevailing directions of displacements in the earthquake foci. It was thought, which is natural, that these directions could be taken as the direction of the predominant tectonic movements within seismofocal zones. However, the results were ambiguous. In the 1950s, Benioff (1954) arrived at the conclusion that there are different types of displacements in seismofocal zones, among which are upthrust, underthrust, longitudinal slip and out-thrust directions. Benioff suggested that upthrust movements (caused by upthrust of the hanging continental wings of seismofocal zones into the lying wings) were predominant. Later, the same author (Benioff, 1959) suggested (as had some other researchers) that the predominant displacements within seismofocal zones of the Pacific surroundings are right-hand slips along the strike, connected with counterclockwise rotation of the Pacific foundation as a whole. More recently, Benioff inferred that the Pacific foundation cannot revolve in such a way due to its complicated configuration (Benioff, 1962).

Also in the 1950s and 1960s, other investigators like V. I. Keilis-Borok and co-authors (Gotsadze *et al.*, 1957) and J. Hodgson (Hodgson, 1962, Hodgson and Stevens, 1964), obtained different results, which were rather close to one another. Keilis-Borok and colleagues, using a sophisticated method of determination of the displacement planes in the earthquake's foci, and employing additional data on propagation of shear waves, concluded that the area of development of recent seismofocal zones in the north-western Pacific exhibits mainly latitudinal displacements along the planes, which correspond to a number of large faults, most often of steep northward dip, but transversal to the strike of island arcs and having mostly large horizontal components of shift. Also dealing with those areas surrounding the Pacific and Indonesia, J. Hodgson found, in accordance with the disposition of nodal planes, two probable alternative directions of displacements — along and across the strike of seismofocal zones and island arcs. In both cases, the dip of most planes of displacements proves to be steep, with the direction of displacements close to the subhorizontal slips. The author admitted that these transverse slips, since they partially correspond to the observed faults, are more plausible as displacement directions, compared with other suggested solutions. Within each of the island arcs, in the author's opinion, left-hand and right-hand slips are displayed equally.

Another way to explore the regularities of displacements in earthquake foci is to investigate the orientation of stresses which, it is supposed, cause the earthquakes. This is a less direct way because the orientation of supposed displacement planes is obtained more directly from data of seismological observations, although these are also not quite reliable (in particular, as indicated above, because of the arbitrary choice between two equally possible solutions). But this approach has an advantage (which seemed to be imaginary) of exchange of two planes of possible displacement for one stress vector, which was situated between these planes, if one uses the most extensively employed mechanical model of rupture which suggests displacements along two shear planes inclined to the vector of stress with an angle of 45°, or close to it. True, in addition to the stress vector, one obtains a tensile vector normal to it which can also be a cause of emergence of the same displacement planes; but, as a result of this exchange, the question of direct judgment concerning tectonic displacements through displacements in earthquake foci turns out to be lost. Nevertheless, determination of stress directions sometimes offers a more homogeneous picture of vectors compared with the determination of displacements in earthquake foci, so these methods have also been widely applied. Many elaborations based on these methods have resulted in concepts of orientation of the stress axes, compression or tension, transverse to the strike of seismofocal zones. However, such orientation, even if it exists, does not mean that in the earthquake foci there are uniform displacements which may be quite different. Furthermore, conclusions about the orientation of stresses prove also to be very varied.

According to H. Honda, A. Masatsuka and M. Ichikawa (1967), in the area of the Japan Islands, for the shallow earthquakes, compressive stresses oriented transverse to the island arc strike and almost horizontal are mainly typical. With intermediate and deep earthquakes, the orientation of horizontal projections of compression vectors across the strike of the seismofocal zone mainly persists, but in most cases they display rather large sloping angles. The orientation of tension axes conjugated with those of compression is more irregular; however, they, too, are partially oriented transverse to the strike of the Japan Islands, especially of the Bonin Arc where they are mostly associated with deep earthquakes, down to 500 km (so the subhorizontal transverse compression can correspond to the underthrusting of the oceanic wing in the seismofocal zone, provided the axes of tension are subvertical and making the appropriate choice of one of two possible planes of a displacement, but the similarly oriented extension corresponds to a reverse movement, i.e. to out-thrusting of the oceanic floor; a random orientation of the rest of the extension axes testifies to a random orientation of displacements).

According to A. Ritsema (Ritsema and Veldkamp, 1960; Ritsema, 1964), the areas of island arcs are characterized by considerably inclined positions of the main axes of tension and compression, with orientation of their horizontal projections mainly transverse to the strike of island arcs. Ritsema classifies all earthquakes into two groups: one is characterized by action in the horizontal direction within the plane of main stresses (tensile — compressive) of the compressive force, and the other of the tensile one. The horizontal compression, if the shear plane is inclined, results in upthrusts and overthrusts, with the extension in normal faults or out-thrusting of lower wings of fault structures. Among the earthquakes studied by Ritsema, those which are associated with extension proved to be most numerous; the author also points out that, according to his evidence from the Indonesian and Philippine Arcs, the horizontal extensions at the depths of 100-200 km are predominant — a finding supported by the Ichikawa evidence from Japan.

L. M. Balakina (1962, 1979, and other publications) is known as an adherent of the viewpoint of predominant distribution of transverse subhorizontally oriented compression in the Pacific margin zones. However, tensile and intermediate stress axes, in her opinion, often change places (subhorizontal orientation of extension axes parallel with the seismofocal zones corresponds to the transversal to the strike of these zones offset along the steeply dipping planes, while their subvertical orientation corresponds to the underthrusting of the oceanic wing towards the dip of the zones, or, in the alternative version, to upthrusting of the same oceanic wing on the continental one). The data obtained by Balakina were used in some works on tectonics as a proof of underthrusting of the oceanic wings of the marginal seismofocal zones, although Balakina herself treats them in quite a different way. 'The results presently available from determination of orientation of stresses in the earthquake foci of the Pacific seismic belt cannot be considered as a proof of the validity of the plate hypothesis since deeper than 100 km, the pattern of stress orientations turns out to be unsteady and changeable from area to area and there are no grounds to assert that the basic feature of stress orientation is its coincidence with the dip in the corresponding focal zone" (Balakina, 1979, p. 30). This conclusion is reinforced by consideration of displacements in the earthquake foci: their varying orientation shows that "the focal zones in the areas of island arcs could hardly be treated as single faults on a planetary scale. Rather, these are the zones where there are a large number of differently-oriented ruptures" (Balakina, 1962, p. 51).

This author pays special attention to earthquakes generating tsunami (Balakina, 1972). They normally occur on the continental slopes of the marginal deep-sea trenches and result in instantaneous deformations of the sea floor which, in some cases after the strongest earthquakes, are immediately detectable. So, during the Alaska earthquake of March 28, 1964, the floor surface rose on the ocean side and descended on the continent side of a longitudinal fault, with an amplitude of several metres; all this occurred near the fault line stretching over hundreds of kilometres along the slope of the Aleutian Trench, with more elevations than descents. In determining the position of nodal surfaces according to the seismological records, it was established that one of them usually strikes along the continental slope with its subvertical or ocean-inclined dip (such a plane with a steep angle of dip towards the ocean has been found in particular for the above-mentioned Aleutian earthquake); the orientation of the alternative nodal planes are rather monotonous and in most cases gently sloping, which is, obviously, less favourable to generation of tsunami. If we assume that the displacements along steeply-dipping surfaces are mostly a real cause of the generation of tsunami (which is convincingly argued for by Balakina), these strongest and most widespread earthquakes clearly contradict, in the direction of their displacements, the subduction scheme and, in particular, an assertion that the displacements of upthrust type prevail at near-surface levels.

However, V. N. Aver'yanova (1965, 1975) argued against Balakina's conclusion on uniform orientation of compressive stresses. She investigated the earthquakes in the Kuril-

Kamchatka zone in great detail and arrived at the conclusion that the stress fields in the seismofocal zone are very heterogeneous, with equal amounts of sizeable volumes of all-sided uniform compression and similar extension within this region. The former originates displacements in the plane of the seismofocal zone which this author describes as upthrust —

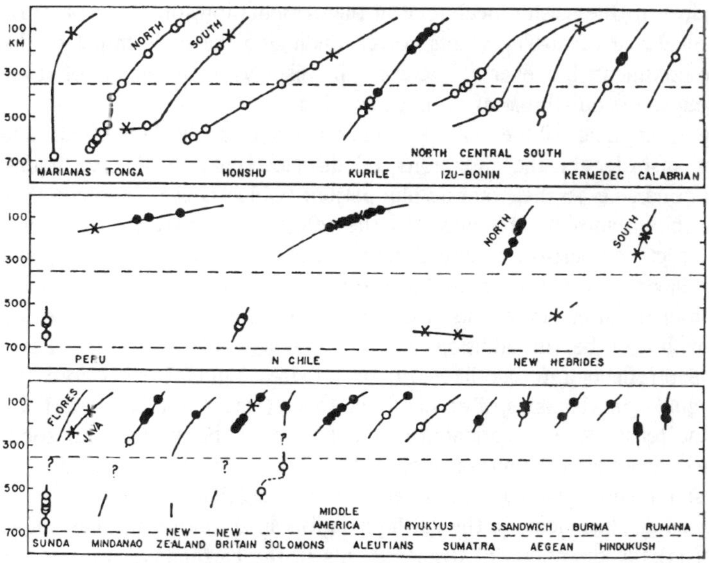

Fig.29. Distribution of tensile and compressive stresses in different seismofocal zones of the world (after Isacks and Molnar, 1971). Circles—main stresses are oriented according to dip of seismofocal zones, black circles—tensional stresses, white circles—compressional stresses; crosses—compressional axes are oriented along a strike of seismofocal zone, tensional axes are subhorizontal.

strike-slip displacements; they may also respond to underthrusting of the oceanic wing. An all-sided extension causes a normal fault — strike-slip displacement which may also respond to an out-thrust of the same wing (the strike-slip constituents act in both cases along the strike of the zones). A complicated pattern of the compression and extension volumes testifies, according to Aver'yanova, to differently directed movements in the Kuril-Kamchatka zone; the magnitude of seismic dislocations due to extension, i.e. those which may correspond to the out-thrust of the oceanic floor, according to Aver'yanova's calculations, exceeds considerably the magnitude of dislocations due to compression. Aver'yanova's observation that the aftershocks of the compression-caused earthquakes normally keep the orientation of the main earthquake, whereas with extension-caused earthquakes, the aftershocks following them are often re-oriented, is worthy of attention. Thus, incomplete data that researchers use overestimate the distribution of compressive stresses. Rather similar data were obtained by I. Kuzin (1974).

Special techniques for studying stresses under the Kamchatka and the Kuril arc were used by Simbireva *et al.* (1976). In their determinations, they rely on data for more than 1700 earthquakes of different intensity, including very weak events (while other researchers usually confine themselves to dozens or some few hundreds of strong and relatively strong earthquakes). The above authors made a statistic generalization of the basic parameters of earthquakes in their foci (the position of the main stress axes, shear planes and vectors of displacements) for some parts of the Kuril-Kamchatka arc with relatively uniform stress fields. According to their observations, the most ordered stresses are those which cause earthquakes of magnitude 7 or stronger; the stresses associated with weaker earthquakes generate a heterogeneous mosaic-like field. These authors identified 14 different types of stress tensors (taking into consideration the ratios of absolute values of main stresses) typical of different areas. Accordingly, the directions of displacements in the Kuril—Kamchatka arc prove to vary significantly from one area to another. According to their data, among the displacements along the planes close to the seismofocal zone in their orientation, the flank parts of the Kuril Arc display a prevalence of displacements corresponding to the underthrust of the oceanic wing towards the continent; however, its central part displays a prevalence of reverse movements corresponding to the out-thrust of the oceanic wing.

With the emergence of the plate tectonics concept in the late 1960s, there was ample evidence of disorder in the directions of displacements in the earthquake foci within seismofocal zones, i.e. to their mosaic-like pattern (which is characteristic of stress directions too, as was shown above). This was why the adherents of plate tectonics did not in general search for direct comparison of directions of displacements in the earthquake foci with the tectonic displacements in the seismofocal zones. They suggested that displacements in the earthquake foci are attendant events which depend on the universal dip-directed orientation of stresses, which can be, in addition, compressive as well as tensile. Essentially, such a conclusion also reflects a mosaic-like structure of stress fields. Furthermore, the investigations of the earthquake mechanisms of seismofocal zones (Isacks and Molnar, 1971, for example) show that the tensile stresses oriented along the dip of the seismofocal zones, at all levels, turned out to prevail over compressive ones (Fig. 29). So this solution cannot yield support for subduction statements, which was considered in more detail in Chapter II.

The data on displacements in the earthquake foci in seismofocal zones allow the following conclusions: (I) the directions of displacements in the earthquake foci are of an apparently disordered pattern and do not exhibit any prevalence of displacements with some definite direction; (2) the directions of stresses determined from the chosen versions of displacement directions also prove to be diverse and of no regular pattern. So no direct determination of direction of tectonic movement in these zones from these data is possible.

Eduction interpretation

Almost all the models considering joint interpretation of seismological and geological materials related to seismofocal zones contain a contradiction. This comprises the determination of the position of the surface to which the main tectonic displacements pertaining to seismofocal zones are attributed. This surface is usually placed among the earthquake foci where they are thickest. In doing so, it is also supposed (or it follows from the model) that the mechanical properties of the environment on both sides of the surface of main displacements are approximately equal. The surface constructed in this way always outcrops onto the Earth's surface amidst the continental slope. Some authors cite such a position for the normal tectonic surface as an established regularity. Meanwhile, a geological approach says that amidst the continental slope there is not any clear boundary which would divide especially heterogeneous structures. Such a boundary exists at the foot of the continental slope near the trench axis and separates the crust of a continental type or a transient type, rather similar to the continental one, which entirely makes up the continental or island arc slope (with occasional exceptions of its foot proper) from the oceanic crust which creates the floor of the trench axis and its outside slope. Nor does the plate tectonics hypothesis itself cast any doubts on such position of the basic division surface—those of the main tectonic movements. In this case, there arises an obvious discrepancy between the position of the same boundary if it is drawn in accordance with the relief and proceeding from the location of the earthquake foci; but it does

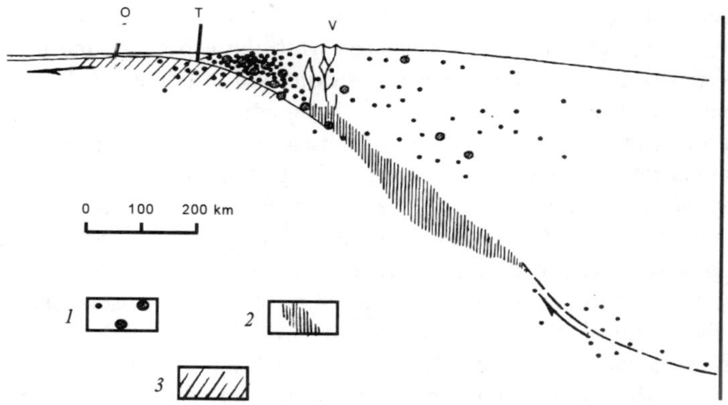

Fig. 30. Interpretation of seismological data from the eductional point of view, generalized section across active continental margin. Direction of outflow of mantle matter along hidden edge of the continental lithosphere is shown with arrows. 1—Earthquake foci; 2—volume of forming of magmatic melts corresponding to zones of low seismic quality; 3—volume of fast cooling of out-thrusted masses and formation of solid oceanic eduction crust. T—trench axis O—outer oceanic swell in front of trench, V—volcanoes at continent's edge. Vertical and horizontal scales are equal.

not attract any attention, and usually is not discussed, possibly because plate tectonics models use very small scales and drawings that are too schematized. Interpretation of seismologic data in accordance with the eduction point of view, first of all, takes away this discrepancy. As already shown, the essence of the eduction process is the outflow of the depth mantle material from the continental edge, which in this case should be attributed to the whole depth of the seismofocal zone with deep consolidation and rigid properties that are much higher if compared with the oceanic lying wing of the seismofocal zone. The grounds for this assertion can be found in Benioff's works and in those of some adherents of plate tectonics (Ringwood, 1975, etc.). If deep earthquakes depend as much on the rigidity of the continental lithosphere as shallow ones, then this lithosphere comprising the continent's edge, which exerts mechanical control over outflow of the hotter and more plastic mantle material, undergoes mechanical effects, such as pressure, within this material: so within a more rigid continental edge, the ruptures and their accompanying earthquakes should be spatially distributed. In this case, the surface of the main tectonic movements should be thought to be located not among the earthquake foci, as is usually done, but along the ocean-facing boundary of the volume of the earthquake distribution (which is, as was mentioned above, normally most pronounced and inclined towards the continent). The surface of the main tectonic movement determined in this way invariably outcrops onto the Earth's surface along the trench axis at the foot of the continental slope, i.e. where it should outcrop (Figs. 3, 4, 5).

The model of the outflowing of mantle matter from beneath the edge of the continental lithosphere is shown in Fig. 30. According to this, the earthquakes of the seismofocal zone are, mostly, a reaction of a relatively rigid hanging continental wing of the eduction system to the movement of the depth material of its lying wing. Orientation of displacements in the foci of the earthquakes which originate in this way should be various — which is, indeed, what is observed in reality. Broadening of seismofocal zones (as volumes of distribution of earthquake foci) when approaching the Earth's surface is accounted for by a growing rigidity in the continental lithosphere. Thickening of the earthquake foci (a specific feature in the structure of seismofocal zones which was pointed out by the author previously, that leads, when the surface drawn among the earthquake foci, to its outcrop onto the middle of the continental slope) is disposed right at the outflow surface which is also natural, but only from its continental, i.e. low-plasticity, side. Association of earthquakes with those parts of the lithosphere of high elastic properties complies with considerations expressed in Chapter I. When we move away from the boundary of outflowing, earthquake foci gradually vanish. In the lying (outflowing) wing, the earthquake foci can only be found in its near-surface part, at a depth less than 100 km, which should be attributed to the cooling of rising matter and loss of its initial plasticity.

As already mentioned, the eduction process is accompanied by a transfer of much heat from the depths to the Earth's surface. At levels of 500-250 km, where the solid phase is mixed with liquid and gaseous ones, due to preservation by the rising matter of anomalously high temperatures and dropping pressure, there are formed zones (`prisms') extended along the surfaces of the outflow, of high absorption and low velocities of seismic waves. There are much fewer earthquake foci here. The zones of reduced velocities of seismic waves are accompanied from below by volumes of anomalously high velocities and high seismic quality (see Fig. 7). Their emergence can be accounted for by the presence in the out-thrusting wing of the components taking no part in the melting process (due to its high melting temperatures, being possibly of the restite type); so this matter is of heightened density for the levels that it occupies, which is also natural under the processes of eduction upwelling.

It seems possible to assert that the above general scheme is in more complete accord with seismologic data than are other schemes, in particular, the subduction one.

GRAVITY ANOMALIES AND RELIEF OF ACTIVE MARGINS

The concept of the out-thrusting of the oceanic floor explains well the emergence of sharply expressed gravity isostatic anomalies typical of the island arc provinces and active continental edges. It is known that island arcs, as well as the mountainous margins of the continents adjacent to the oceanic trenches, are accompanied by large linear positive isostatic anomalies, the axes of trenches with similar but greater negative anomalies, and, finally, the oceanic swells beyond the outer slope of the trenches are once again accompanied by positive anomalies (see also Chapter I). With the out-thrusting of the oceanic floor, the above gravity anomalies, as well as the relief of island arcs and oceanic trenches, turn out to be a natural consequence of dynamic conditions in the area where the continental lithosphere is ruptured (Fig. 31).

The origin of the basic relief features in the active marginal provinces can be explained in the following way.

1. The mantle masses rising from depth should strive, near the site of their arrival at the surface, to raise up above the equilibrium isostatic level the edge of the continental lithosphere over those mantle masses, which are adjacent to the trench as an active continental margin or its detached part in the form of an island arc. This results in a mountainous relief of such continental margin and island arcs; as these mountainous structures were raised by endogenic tectonic forces, together with their underlying mantle material, to a position above their natural isostatic level, so they are accompanied by positive gravitational anomalies.

2. At the site where mantle masses come to the Earth's surface to become new oceanic crust, they should always occupy a place lower than their natural isostatic level if they were oceanic crust free of any load, since they yet bear the load of the neighbouring edge of the continental lithosphere. When relieved of this load behind the line of the continental lithosphere rupture, the oceanic crust should experience the effect of the vertical isostatic forces striving to

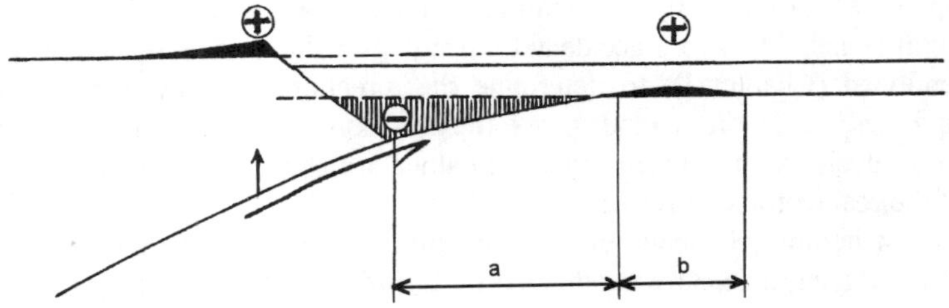

Fig. 31. Origin of positive and negative isostatic gravity anomalies near active ocean — continent or ocean — island-arc margins. Dot-and-dash line — level of natural isostatic equilibrium of continental lithosphere's surface (simplified), dotted line — same, of oceanic crust surface. Black — masses lifted above natural isostatic level, vertical hatching — space with insufficient mass below natural isostatic level; a — area where isostatic equilibrium is gradually reached due to action of upward isostatic forces (outer slope of trench); b — area where lifting isostatic forces stop acting and temporary piling of material (as oceanic swell) occurs. Arrows show direction of movement (out-thrusting) of depth matter and direction of vertical pressure force exerting upon edge of continental lithosphere.

raise it to the level of the isostatic equilibrium. Displacement of the crust up to this level cannot take place instantaneously, since the oceanic crust comes to the surface in a rather cooled-off and rigid state; the level of the isostatic equilibrium is attained by a gradual ascending of the oceanic floor with its oceanward movement. In this process, nothing can form except for a typical oceanic trench with a steep and high inner slope (which is the rupture of the continental lithosphere) and more gentle sloping and lower outer one (the surface of a gradually rising oceanic crust). The distance within which the oceanic crust reaches the level of the isostatic equilibrium, i.e. the width of the trench outer slope, depends on the strength properties of the out-thrusted oceanic eduction crust and should be about equal everywhere — which is indeed observed in reality. Throughout the distance where the oceanic crust ascends, there is everywhere mass deficit which causes a negative gravitational anomaly, which is at its greatest near the trench axis.

3. Within the outer trench slope, i.e. at a distance where isostatic equilibrium is being gradually established, the Earth's expansion forces, causing the depth material outflow and oceanic crust out-thrust, are supplemented with upward isostatic forces. These supplementary forces should give, within the outer slope of the trench, a supplementary velocity to the ocean crust travelling oceanwards, compared with the velocity of its further movement within the open ocean, where isostatic forces no longer act. Where the oceanic crust, as a result of reduction of velocity, reaches a state of isostatic equilibrium, there arise conditions for a piling up of material caused by local conditions of compression that are generated here. This material forms an outer oceanic swell rising over the level of isostatic equilibrium of the oceanic crust, which in turn cause zones of positive gravitational anomalies.

GEOLOGICAL EDUCTION MODEL

Seismologic, gravimetric and thermometric data can be successfully coordinated with the eduction viewpoint and, as a result, we have the possibility of explaining each phenomenon as a direct consequence of the eduction process, without elaborating additional hypotheses for each case. The same should be said about explaining the macro-relief of the above-seismofocal areas. The problems of eductional interpretation of geological features are more complicated in many respects since the relevant data are obtained from direct observations and are not subject, unlike geophysical information, to quantitative unification. However, it is more concrete, which sometimes helps to avoid accidental averaging in the models of natural processes. As already mentioned (Chapter I), to determine the directions of movements connected with seismofocal zones, i.e. to prove or disprove the eduction or subduction viewpoint, geological methods are of decisive importance, since they alone are able to unambiguously determine the direction of the real tectonic movements.

The main structural elements of the eduction system are represented by a hanging continental and a lying oceanic wing divided by a surface of the main tectonic displacements, i.e. by the eduction surface proper. The continental and oceanic wings differ greatly in their mechanical properties. A continental wing, if the depth of the seismofocal zone is great enough, is made up of the continental, platform type lithosphere, in the typical case, consolidated down to a depth of hundreds of kilometres. Down to this depth it preserves its elasticity and, as a consequence has, on the whole, considerable rigidity (since the latter, according to its rigorous physical definition, is the product of an elastic modulus by the cross-section of the body). The eduction process consists in an outflow of considerably-plastic depth material along the lower side of the continental wing, which is the eduction surface. Where it outcrops onto the Earth's surface one can find an oceanic trench. A seismofocal zone proper is a volume of distribution of destruction centres which emerge within the rigid continental wing. The latter can also be heterogeneous and consist only partially of old continental lithosphere with the younger lithosphere of folding type of different ages; and if, within this wing, a marginal sea (of a thrust-apart geosynclinal region) has opened, it also consists of a newly-formed oceanic-type lithosphere. In such conditions, volumes of distribution of earthquake foci, i.e. seismofocal zones, can enclose all the above-mentioned types of lithosphere, among which most favourable for the appearance of earthquake foci is continental lithosphere of platform type. In its predominant part, the continental wing can be considered as motionless within the eduction system.

The oceanic wing, on the other hand, is mobile. Its constituent depth material is of temperatures that are anomalously high for the new levels in which it is now situated, and this provides, probably together with other factors, its heightened plasticity. When moving from deep levels to the surface, this material undergoes considerable transformations, which may involve a separation of the most volatile compounds, partial or full melting, recrystallization and, before outcropping onto the surface, diaphthoresis. Once arrived at the Earth's surface, the matter of an oceanic wing is cooled and hardened. Since the matter of the continental and of the oceanic wings differ in their composition and density, the heavier matter of the oceanic wing always takes a lower hypsometric position on the Earth's surface compared with that of the continental wing, and therefore makes the floor of sea and oceanic basins. The process of the out-thrusting of the oceanic wing matter leads to the growth of their foundation and to their widening.

The reason for eduction outflow of depth material should be related to the existence of depth pressure forcing the mantle material to the surface and overcoming the gravity force effect. Such conditions may exist when the Earth's interior expands, which is supported by many other facts.

The eduction process acting in a direction opposite to gravitation is evidence of its great power. There is a profound tectonic rebuilding of the Earth's crust occurring over the eduction zones whose essence corresponds largely to the ideas of tectono-magmatic activation. The same restructuring leads to the emergence of typical features of macro-relief when above seismofocal zones there lie either island arcs with marginal seas behind them (as, for example, in eastern Asia, or, as on the recent Pacific coast of South America, or marginal continental uplifts — abruptly elevated edges of the continents without thrust-apart zones but with marginal seas behind them (which is the Andean type of marginal province). The emergence of marginal continental uplifts, oceanic trenches and outer oceanic swells, as well as their accompanying isostatic gravitational anomalies, is a direct consequence of eduction out-thrusting. Another equally direct consequence is detachment of the foundation of island arcs from the edges of the continents.

The eduction mechanism seems to be better than any other tectonic construction in explaining the emergence of ore-magmatic belts above seismofocal zones. During eduction, the rise (outflow) of mantle material starts at depths of 600—700 km (which is the level of the deepest earthquake foci in seismofocal zones), at temperatures of the order of 2000°C, and at a lithostatic pressure of over 200 kbars. With eduction out-thrust, this material moves consistently to higher levels of lower lithostatic pressure while retaining a temperature anomalously high for these levels. In these conditions, starting from certain depths, an unavoidable consequence is the separation of the mantle material's most fluid components and, therefore, first of all, its degassing. Overheated gases (fluids unable to turn into liquid because of their extra-critical temperatures) seem to be the main agents of transformation of the overlying masses in the continental wing owing to their ability to travel over any distances in the direction of lower pressure. Fluid effects should be associated with development of chambers, caverns, and channels which act as paths for distribution of the fluids themselves in the continental lithosphere above the eduction out-thrust surface, and for transfer of depth heat, as well as for delivery at the upper levels of the Earth's crust of the material of mantle origin, including ore material.

The volcanic belts occupy only a part of the area within which the seismofocal zone is projected on the Earth's surface. In a band of a similar width or even wider, which is more remote from the trench and is disposed behind the volcanic belt, there are numerous intrusions whose age is close to the age of volcanoes in the volcanic belt, and there are also numerous endogenic ore deposits which are also approximately of the same age. Within the area of the Mesozoic margin of the Paleo-Asiatic continent in north-eastern Russia, such a band is often called the `perivolcanic zone', although its width is no less than that of the volcanic band which is usually called a belt (therefore, it would be reasonable to call both bands `zones' within a single above-seismofocal magmatic, or ore-magmatic, `belt'). Young analogies to such plutonic above-seismofocal zones (or belts?), including Pliocene-Quaternary ones, can be found in the Rocky Mountains of North America and in the east of the Andes in South America. In the latter case, a plutonic (intrusive) band covering, in particular, Bolivian ore localities, whose intrusive stocks and mineralization is of a Miocene-Pliocene age, is situated above the present-day active seismofocal zone.

An essential condition for creation, and magmatic and metamorphic transformation of the hanging wings of eduction systems is a broad development within them, both in volcanic and plutonic zones, of depth magmatic channels (translithospheric pipes, magmatic columns, etc.). An example of such formations are channels penetrating above-seismofocal space which can be detected beneath a number of large volcanic centres. The eduction process provides a delivery of much extra heat energy as compared with that required for magmatic melts, and this heat is sufficient to create a network of such channels, as well as generating other additional processes.

Tectonic deformations in an eduction process

A mighty process of eduction out-thrust of depth material is impossible without great tectonic deformations. We have already dwelt on the tensile deformations which result in such large-scale phenomena as the opening of marginal seas and the separation of the detached pieces of island arcs from the continents. However, when proceeding from an eduction model, one could also speak of other kinds of tectonic deformations which develop due to the special conditions caused by eduction, which are essentially compressive. A common feature in such deformations is their dependence on direct or oblique interaction of two power factors — outflowing forces directed from depth to surface and gravity forces opposing them. We shall consider below the main tensile and compressive types of deformations.

Tensile deformations and structures. From the point of view of eduction, the shape of island arcs, convex towards the ocean, resulted from an outflow of the depth material to the surface and indicates its direction. The chains of island arcs look as if they sag towards the ocean while remaining rather closely connected with the continents on the edges. Indeed, the apexes of the re-entrant angles between the arcs are portions of the arcs nearest to the continent where they come together, either immediately (the eastern flank of the Aleutian Arc, and almost so in the south of Japan), or through the continent-adjoining islands or peninsulas (Taiwan, Sakhalin, Kamchatka). The arcs are thus detached fragments of the continents which are fully or partially separated.

The eduction approach also explains the variety of shapes to be found in island arcs. The arcs can look both like narrow island chains or like wider bands of islands which are rather bent, like the Japan Arc, or of more straight form like New Zealand, or may be archipelagoes of more complex forms like that of the Philippines. The shapes of island-arc detached continental fragments may depend on, on the one hand, the inside structure of detached masses and, on the other, on conditions of outflow of depth material, as all outflowing masses create a system of one-sided mantle currents directed to the Earth's surface with their own peculiarities.

As an example of a large detached mass bearing the remnants of an ancient day surface which is, however, submerged to a considerable depth below sea level, one can consider the Obrutschev Rise in the north-western Pacific, when approaching this eductionally. According to the dredging results, described by B. I. Vasilyev and G. B. Udinstev (1982), on this rise, the outcrops of bedrocks of various composition are distributed and they are compared by the authors with different-age complexes of the Central Kamchatka—from the metamorphic Ganaly Series to Meso-Cenozoic volcanogenic covers. An overlying sedimentary cover is of subhorizontal occurrence and total thickness of 1044 m; it is represented, according to the deep sea drilling of *Glomar Challenger* (Site 192), by shallow calcareous sediments of the Maastrichtian, Eocene and, possibly, Oligocene age in its lower part; by terrigene, largely clayey sediments of high rate of accumulation in its middle part; and by middle Miocene biogenic diatomaceous oozes of low rate of accumulation in the upper part. Thus, shallow sediments from bottom to top over the section are replaced with more abyssal ones rather similar to the facies of continental slopes, and then with typical abyssal sediments. The thickness of the sedimentary cover of the Obrutschev Rise as a whole is much greater than that of sediments in adjacent trenches, which in the Aleutian Trench is as little as 200 m (according to the data of the above-mentioned authors), while in the Kuril Trench it is 500-600 m, occasionally more, according to the seismic investigation. Vasilyev and Udintsev arrived at a conclusion that the foundation of the Obrutschev Rise is of a continental nature and originally was structurally united with the Central Kamchatka but later, after the Kuril-Kamchatka and Aleutian Trench were laid down in the late Miocene — early Pliocene, it separated from

Kamchatka; subduction in these trenches is out of the question, according to the authors' conclusion.

Although Vasilyev and Udintsev do not consider the separation of the Obrutschev Rise from the Central Kamchatka as a consequence of eduction, an eduction explanation of the events which they described seems most appropriate. Indeed, having separated from Kamchatka, the Obrutschev Rise must also have moved apart from it (which, in particular, follows from the ideas of many authors concerning the trenches belonging to extension structures). Changes with time in the facies of the sedimentary cover of the Obrutschev Rise clearly indicate its consistent removal from the coast and the sources of delivery of terrigenic material. The small thickness of sediments on the floor of the neighbouring trenches provides evidence for a rather late origin of these sediments and of the floor too.

Yu. A. Kosygin (1982), who disagrees with the views of the author on the possibility of eduction, refers to a probable absence of the oceanic basement in the composition of foundation of the Obrutschev Rise. Kosygin also has a different interpretation of the data of Vasilyev and Udintsev on the continental-type rocks distributed there. However, according to the above explanations, the oceanic foundation of eduction origin should exist in that case only between the continental slope of Kamchatka and the Obrutschev Rise and between the latter and the island slope of the Aleutian arc.

Out-thrust deformations. The consequence of the out-thrusting of the oceanic wings of eduction zones is distribution in the eduction oceanic basement and in its adjacent continental crust of tectonic slabs, wedges, scales, etc., being oriented along the surface of the main tectonic displacements. Such formations are widespread over many geosynclines, recent and ancient, but in most cases they are treated as a consequence of overthrusting (because of their dip from the ocean). However, it can be shown that in many cases, the age alternation of rocks in the slabs is the reverse of that needed for overthrusts (with their becoming steadily younger in the upper slabs instead of becoming older).

Transversal shifts in above-seismofocal areas. So tensile deformations in above-seismofocal areas accompanying the eduction process are of a large size and specific nature; their essence is in the tension and tearing up of the lithosphere. These can be quite naturally supplemented deformations which are transverse to the trend of seismofocal zones. Such deformations are indeed observed (in particular, they are reflected in the directions of the displacements determined for some earthquake foci, as has been shown above) and vary from relatively small, caused by the horizontal oceanward bend of island arcs during eduction mantle outflow, to comparable in their size with the size of the opening of marginal seas and directly resulting from this opening. In Fig. 32, as an example, is represented a scheme of such displacements along the Sakhalin Island, mainly on the grounds of numerous data contained in the works by S. V. Rozhdestvensky (1975, 1982, and many others), as it was displayed in the Russian edition of this book, but with a designation of the place where the Neftegorsk earthquake of the October 1995 in Sakhalin occurred. According to my interpretation, as was shown in this diagram, the submeridional, all right-sided, horizontal displacements along faults in Sakhalin are due to tension connected with the bending of the Kuril arc and the opening behind it of the abyssal South-Okhotsk Basin. Indeed, the catastrophic Neftegorsk earthquake, with many human victims, was accompanied by an instantaneous right-sided horizontal displacement up to 7 m in amplitude, along the nighbouring submeridional fault (Rogozhin, 1996). These geological notions, however, were little-known, so this displacement was unexpected to many seismologists.

Since the width of the opening of marginal seas reaches hundreds and even thousands of kilometres, the amplitude of their accompanying horizontal displacements along fault zones, strictly transverse and oblique to the strike of the island arcs, can also amount to hundreds of kilometres; this means that these horizontal shifts are among the largest on Earth. Some possible examples of such faults are cited below in the description of the ancient active margins of Asia within north-eastern Russia.

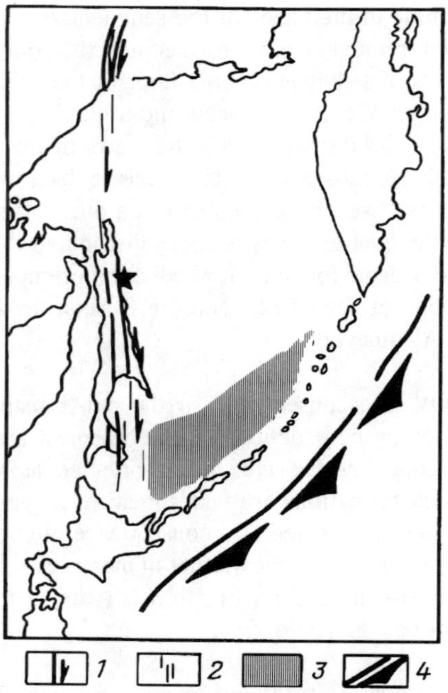

Fig. 32. Right-sided horizontal displacements along longitudinal submeridional faults on Sakhalin Island (after V. S. Rozhdestvensky, 1982, supplemented by the author) as a consequence of large-scale deformations associated with the life of the Kuril eduction zone. 1 — Right-hand horizontal displacements in eastern Sakhalin, 2 — their likely continuation at sea floor; 3 — abyssal South-Okhotsk Basin, 4 — Kuril-Kamchatka trench axis, arrows showing direction of lying wing out-thrusting in eduction zone. The place of the Neftegorsk earthquake (accompanied by instantaneous horizontal slip along submeridional fault up to 7 m in amplitude) is shown as a triangle.

Folding and other deformations of lateral compression. The outflowing material of the lying wing of eduction zones raises the edge of the hanging continental wing above its natural isostatic level which is an obstacle to the outflowing, and this is a cause for mountainous rises in continental edges and island arcs which have become in disequilibrium isostatically. However, with such a rise, there must exist another constituent of the main force which is normal to the surface of mantle outflow and arising from the pressure delivered by the outflowing lying wing of the eduction system to the continental hanging wing (Fig. 33). If the seismofocal zone is inclined at an angle of the order of 45° (which is quite common), this other constituent is about equal to the vertical constituent. One could do a simple calculation which would show that such a horizontal constituent is enough to cause the rupture deformations in

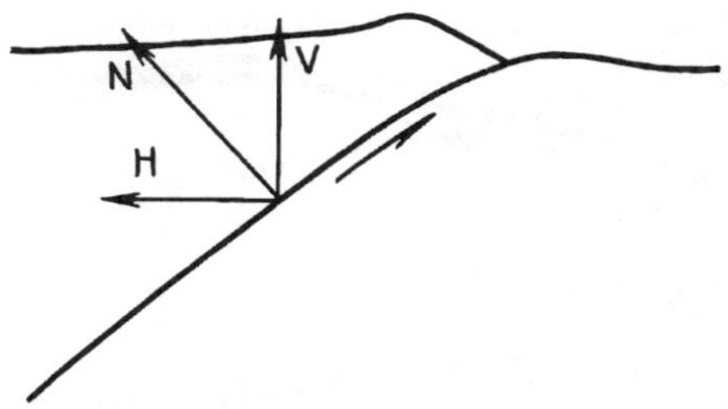

Fig. 33. Direction of main forces in eduction process. N — normal force acting on hanging wing and caused by upward pressure of outflowing material, V and H — its vertical and horizontal constituents. One-sided arrow — direction of outflow of depth matter causing both upward pressure and extension experienced by hanging wing of eduction system.

the hanging wing of an eduction zone, the more so folding, if this wing contains portions of the oceanic (geosyncline) lithosphere of reduced thickness and, consequently, reduced rigidity compared with the ancient continental lithosphere. Let us assume that a width of a marginal rise or of the foundation of an island arc makes up 250 km and an average height of their elevation above the isostatic level of 3 km. In this case, the vertical force acting on the element of a cross-section in a hanging wing of 1 cm width will make up about 2×10^{10} kg. If the same force acts in a horizontal direction to a hanging continental wing and within the latter there is, behind a marginal rise, a portion of a geosyncline type lithosphere of a thickness of, say, 100 km, then horizontal compressive stress in this portion will reach 2000 kg/cm^2, i.e. it proves to be destructive. So that part of a hanging wing will undergo compressive deformations described as a whole as folding, which embrace not only the sediments but the whole lithosphere section. The other parts of the hanging wing made up of an ancient continental lithosphere, which are consolidated to a depth of several hundreds of kilometres and have several times higher rigidity (the product of an elasticity modulus by lithosphere's section) can withstand the effects of the same force. As follows from this discussion, lateral compressive deformations are inherent in hanging wings of eduction zones only; in the next chapter there are a series of examples of this rule.

Since the action of a lying wing on the hanging one consists in not only upward pressure but also in extension conveyed by their friction (from the movement of the hanging wing towards the Earth's surface), the results of combination of tension and compression in the hanging wings of eduction systems can be various.

Overthrusts in the rear of marginal rises. Elevation of continental rises and island arcs higher than a natural isostatic level causes an unsteady dynamic situation wherein a gravity force tends to destroy the rises. In combination with a horizontal compression caused by an eduction outflow, this leads to a gravitational slide and other displacements of masses from elevated portions of marginal rises to their rear areas where unbalanced conditions dwindle to nothing. Overthrust deformations of such a character and direction are known, for example, at the foot of the eastern slope in the Andes and are also found even more often in north-eastern Russia to the north-west of the Okhotsk-Chukotka volcanic belt. The overthrust (and inthrust) deformations of such a type are obviously widespread over the entire Pacific surround, in particular, on the continent-facing slopes of island arcs.

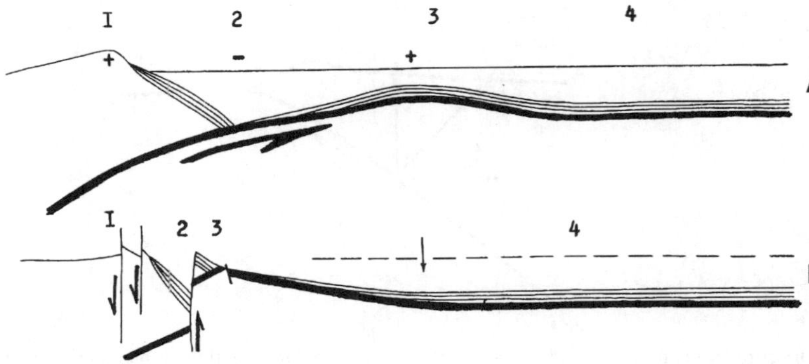

Fig. 34. Occurrence of different-age sedimentary beds within geosynclines or parts of oceans adjacent to trenches if they are created by eduction. A—sediments in trench and adjacent basin as they distributed as a result of the eduction process; B—same after eduction activities have ceased and isostatic vertical compensatory displacements have occurred.

Vertical compensatory deformations. I have dwelt above on the explanation of gravitational isostatic anomalies which develop at the site of trenches and on its sides as a consequence of tectonic movements in the active marginal areas. It is natural that when the tectonic movements stop (this is true of any method for explanation of isostatic anomalies) the isostatic forces counterbalanced by the previously tectonic movements will prove to be solely acting and, sooner or later (after the lithosphere strength is overcome and it gets destroyed over the parts of maximum distortions of isostatic balance), will cause vertical compensatory displacements restoring isostatic equilibrium. The axial zone of the trench which was characterized by a mass deficit will be raised up to its natural isostatic level, and the zones of positive anomalies, i.e. island arcs and continental marginal rises, as well as the oceanic swells on the oceanic side of the trenches undergo descent. Thus there will occur a kind of inversion of structures which existed previously during the active stage of the seismofocal zone, with preservation in the shapes of structures of their inherent linear forms and predominantly subvertical new displacements (Fig. 34). It is obvious that such deformations are most typical of later stages in development of marginal areas and correspond to the time of extinction of eduction zones. Most pronounced should be horst-like rises in the place of the trench axial parts because of their greatest absolute values of anomalies. Formation of such compensatory rises in the place of trenches is not accidental but a regular phenomenon which has so far not been adequately taken into consideration in the analysis of development of tectonic structures in the marginal areas.

Evidence of eduction

The geological model suggested allows us to identify geological signs which are characteristic of the eduction process. Three types of evidence seem to be unequivocally characteristic, corresponding only to the eduction process among all possible.

Areal distribution of a tectonized cover (melange) over the surface of the oceanic (geosynclinal) basement. Over dozens of kilometres of its path before emerging onto the Earth's surface, the out-thrusting wing in the eduction zone, as was noted above, continuously interacts with material of a stationary continental wing, destroying it while being destroyed itself. This phenomenon is related to solidification (due to cooling off) of the out-thrusting wing material which was plastic deeper than 100 km (one can estimate this from the distribution of the earthquake foci, as only higher than this depth they can be found not only in the hanging but also in the lying wing of the surface of the main displacements attendant by the seismofocal zone). Thus, an inevitable consequence of eduction out-thrusting is destruction of rocks along the zone of main movement. With an amplitude of displacement of dozens or even some few hundreds of kilometres, the thickness of the destruction zone should be of several kilometres. So, during eduction out-thrusting, some products of tectonic destruction should be brought onto the Earth's surface; consequently, the oceanic (or geosynclinal) crust of eduction origin should bear in its upper part, which is directly overlain with sediments, a thick cover of tectonized material mainly of mafic, basic and ultrabasic, composition corresponding to the mantle and continental crust material (the latter essentially from the lower parts of the crust). This cover, according to this argument, should partly include also a material of the continental granite layer of the crust too: thus it corresponds as a whole to the well-known formation of melange. So the wide distribution of melange in the composition of oceanic (geosynclinal) basement should be considered evidence of the eduction tectonic process.

Age changes of the sedimentary cover base. During the eduction opening of the geosyncline (of marginal sea type, for instance) it becomes asymmetric in structure, which is reflected, in particular, in the composition of its sedimentary cover. As the eduction out-thrust is on one side of the geosyncline, or the ocean, the age of the eduction basement and, correspondingly, of the sedimentary cover base becomes more ancient farther from the trench, while the age range of sediments in the cover increases (Fig 34, A). As to the trench proper, as long as the eduction mechanism acts, only the youngest sediments are present there (as can be observed in recent trenches). The sediments' facies which deposit in the trenches and are characteristic of them, are transported oceanward and, during such a process, are distributed over the entire geosyncline area but become increasingly older farther from the trench. The structure of the sedimentary cover proves, thus, to be similar to the spreading one but of a reverse direction of changes (if one assumes the viewpoint of the new global tectonics) with respect to the trench.

Development of folding only in hanging wings of the acting seismofocal zones. Distribution of forces, acting on the hanging wing of eduction systems, is unstable and can lead to either the extension of this wing, resulting in the opening of the marginal seas and emergence of other tensile structures, or to development of compressive deformations which embrace, in the first place, the basement of the same previously opened marginal seas due to its little thickness and consolidation. The folding, thus, does not necessarily accompany the eduction process, but should it develop, it does so in the hanging wing of the eduction zone; no simultaneous folding or other compressive deformations take place in the oceanic wing, within the adjacent marginal seas or the ocean margins: they continue to expand through eduction while the eduction mechanism, accompanied by the seismofocal zone, acts. The position of the folded areas with respect to the acting seismofocal zone again becomes the reverse of the one implied by the subduction model, according to which the folding-causing forces are always directed from the ocean; so the oceanic and marginal seas' basement should undergo folding too (an assumption on the `absolute' rigidity of these basement is out of the question as, without fail, it will be folded in due time). In practice, when the position of the paleoseismofocal zone is established, the appropriate estimation of the place and time of the folding is fully met with the possibilities of the geological mapping.

The two above-mentioned bodies of evidence can be used to identify the eduction process (and a selection between the eduction and subduction model), both in the ancient geosyncline provinces which experienced folding, when carrying out geological explorations and mapping, and over the recent marginal seas and parts of oceans adjacent to the trenches if one employs the deep-sea drilling data. The third piece of evidence is mainly applicable in investigations carried out in ancient folding provinces, for which the existence of paleoseismofocal zones with accompanying opening of geosynclines is established. An estimation of the presence of a tectonized foundation and of increase of the age of the sedimentary base farther from the trench can be unambiguous if the investigation is thorough enough; therefore, a geological solution of the question of whether an eduction process is present or absent is decisive.

Additional evidence of eduction, as mentioned above, is the continentward inclination of tectonic slabs near the front of marginal continental rises and island arcs, with younger rocks in the slabs towards their inclination, and the overthrusts in the rear of these rises directed to the inside of the continent, and others.

Chapter IV. The Mesozoic Active Margin of North-East Asia

Three most easily identifiable structural zones exist over the territory of north-eastern Russia. These are the province of the Koryak geosyncline which is characterized, first of all, by a distribution of the Upper Jurassic and Cretaceous sea sediments; the Okhotsk-Chukotka volcanic belt to the north-west extending over more than 3000 km, which is a long and wide band where predominantly sub-aerial volcanic rocks cover Mesozoic, mainly Cretaceous in age; and, finally, a vast area farther to the north-west of the volcanic belt which was largely land during the Cretaceous Period but which had had a complex and versatile pre-Cretaceous history. The Okhotsk-Chukotka volcanic belt was, thus, a margin of paleocontinent during the Cretaceous Period. Behind it, there was an area of essentially intrusive activity in the Mesozoic which is sometimes called in the Russian North-East the `perivolcanic zone'.

The majority of tectonists who have recently worked in the Russian North East hold the viewpoint that the border between the Koryak geosyncline province and the volcanic belt with its rear area in the Mesozoic was represented by an ancient seismofocal zone similar to the recent seismofocal zones in the Pacific surroundings. The existence of such a zone was suggested by a number of investigators, among them Shilo and Umitbayev (1977), Belyi (1977), Parfenov (1983) and others. The paleoseismofocal zone was identified mainly on the basis of analogies which one draws between the disposition of volcanic belts in the recent Pacific surroundings and the Okhotsk-Chukotka belt in the Mesozoic. In paleogeographic reconstructions, one determines the association of the active margin province which accompanies the Mesozoic seismofocal zone with the Andes type having no typical marginal seas behind the outcrop of the seismofocal zone on the Earth's surface (Parfenov 1983). On the basis of the position of this paleoseismofocal zone on the border of the Okhotsk-Chukotka belt, it can also be called Okhotsk-Chukotka.

Under the volcanic covers of the Okhotsk-Chukotka belt, a heterogeneous foundation is distributed whose composition can be judged from its outcrops behind the rear borders of distribution of effusive covers and outcrops amidst them. In the composition of the foundation it is possible to identify fairly large blocks of the ancient Archean-Early Proterozoic continental lithosphere which form the Okhotsk, Omolon and Eskimo massifs and, separating them, folded structures of the Early-Middle Paleozoic, Late Paleozoic and Early Mesozoic age, which are characterized by trends transverse to the extent of the Okhotsk-Chukotka belt. The northwest border of the Koryak geosynclinal area with its mainly northeastern common strike of folding structures thus transverses, in a sharply nonconformal way, a number of large structures of the foundation beneath the Okhotsk-Chukotka volcanic belt and of the area adjacent to it in the northwest.

It was mentioned above that a characteristic of the Koryak geosyncline is the distribution of largely deformed sediments of the end of Jurassic and of Cretaceous, and in the southeastern part, also of Paleogene. These sediments form, on the whole, a general sedimentary range reflecting the main stages in development of the geosyncline. The geosynclinal formations of the Koryak territory are sometimes considered to include also the Middle-Upper Paleozoic and partially Triassic deposits distributed mainly over the edges (northwestern and southeastern) of the geosynclinal area. However, these deposits do not make up a continuous stratigraphic section similar to the Upper Jurassic—Cretaceous—Paleogene, but are of a sporadic local distribution making up detached tectonic structures with traces of very complicated displacements. In the author's view, which is substantiated below, these deposits should be

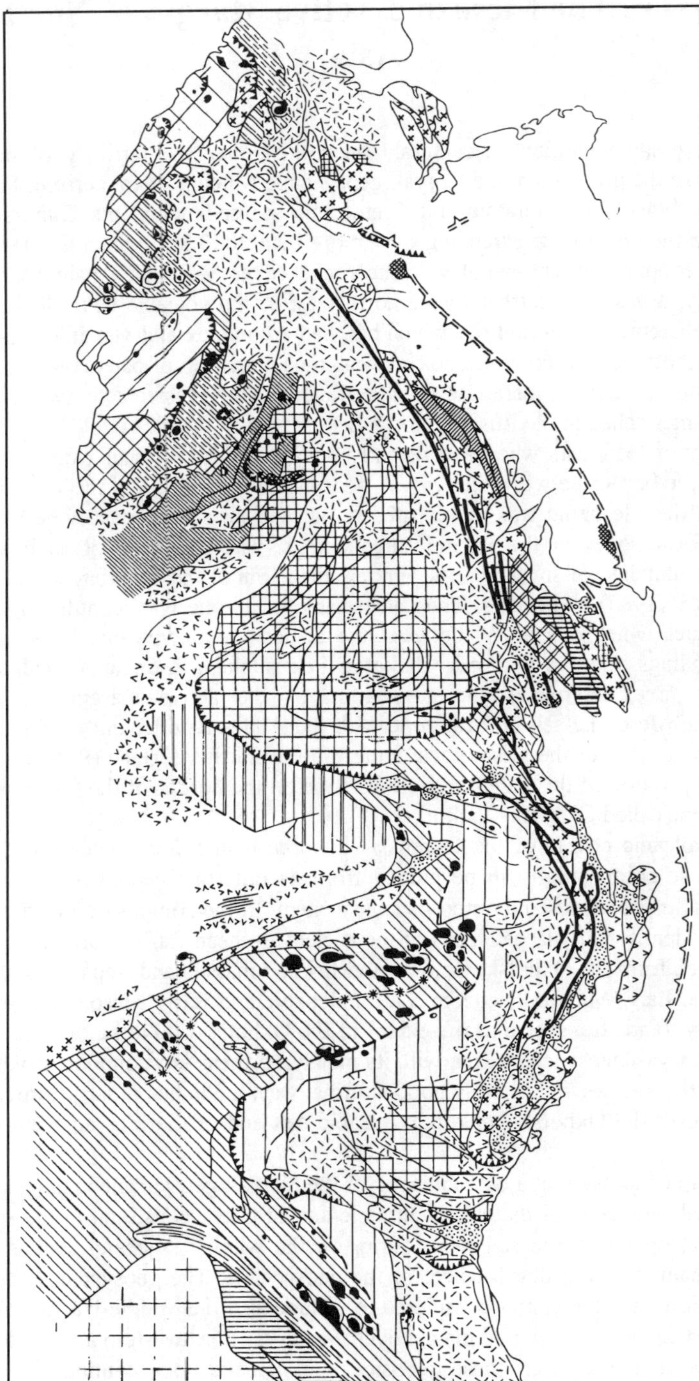

Fig. 35. Structural map of above-seismofocal area of Mesozoic Okhotsk-Chukotka paleoseismofocal zone. 1-5—Structures superimposed on the foundation of the Jurassic-Cretaceous belt of tectonomagmatic activization and its overlying formations: 1—tectonic depressions filled with Neogene-Quaternary deposits, 2—grabens on borders of heterogeneous elements of belt's foundation filled with Aptian-Albian deposits, 3—grabens filled with Upper Senonian—Lower Paleogene basalts, 4—volcanic covers of Mesozoic (Aptian-Cenomanian, partly Neocomian and Senonian) Okhotsk-Chukotka belt and approximate boundaries of their distribution areas, 5—volcanic covers of Late Jurassic—Neocomian age associated mainly with raised pre-Jurassic blocks. 6-12—Paleocontinental slope of Jurassic-Early Cretaceous age: 6—line of outcrop on ancient Earth's surface of the surface of main tectonic movements of Jurassic—Early Cretaceous paleoseismofocal zone; 7—group of large faults separating paleocontinental slope from paleocontinental marginal rise bearing landlocked volcanic belt; 8—deposits on the Mesozoic paleocontinental slope: a—submarine Jurassic—Neocomian sedimentary terrigenic deposits, b—mainly submarine Upper Jurassic—Low Cretaceous intermedium and basic volcanites adjacent to landlocked volcanic belt; 9—Jurassic sedimentary-volcanogenic deposits within thrust-apart geosynclinal structure of paleocontinental slope in Taigonos Peninsula; 10—granitoid intrusions of batholite type distributed predominantly within the paelocontinental slope, including pre-Jurassic ones possibly connected with the slope's foundation; 11-12—outcrops of foundation of Mesozoic paleocontinental slope, correspondingly of Archean-Proterozoic and Middle Paleozoic formations. 13-18—Main structural elements in foundation of Okhotsk-Chukotka belt of above-seismofocal tectono-magmatic activization: 13—Siberian Platform, 14—massifs of ancient Archean and Early Proterozoic platform lithosphere (Okhotsk and Omolon) originally connected with Siberian Platform, 15—part of Omolon massif overlain with Mesozoic and Paleozoic sediments of large thickness and blocks torn away from it, 16—ancient massifs of Hyperborean group (Alyarmaut and Kuyul), 17—extension of Hyperborean massifs under overlying deposits of large thickness, a—of Paleozoic, b—of Mesozoic age, 18—detached blocks of Hyperborean group within the Paleozoic—Early Mesozoic Oloi-Anyui geosyncline system. 19-22—Continental slopes of pre-Jurassic geosynclines: 19—of Early and Middle Paleozoic age, 20—of Late Paleozoic and Early Mesozoic age, 21—of Early Mesozoic (Triassic) age, 22—supposed position of foot of submerged Late Paleozoic—Early Mesozoic paleocontinental slope (roughly coincides with the western boundary of Inyali-Debin trough) of heterogeneous Kolyma massif comprising fragments of ancient platform and folded Paleozoic lithosphere. 23-27—Areas of geosynclines having undergone folding: 23—of Late Proterozoic-Early Paleozoic age (Yudoma-Maya folding system), 24—of Early—Middle Paleozoic age (Sette-Daban folding system), 25—of Middle—Late Paleozoic—Early Mesozoic age (Oloi-Anyui folding system), 26—of Late Paleozoic-Early Mesozoic age (Verkhoyansk folding system), 27—of Early Mesozoic age (North-Anyui—Chukotka folding system). 28-31—Tectono-magmatic structures of central type: 28—intrusion of round section in plan of different, predominantly granitoid, composition and their accompanying ring fault systems, 29—large subvolcanic intrusions, 30—some large tectono-volcanic structures within Okhotsk-Chukotka volcanic belt, 31—ultrabasite massifs (large). 32-33—Some tectonic boundaries: 32—large overthrusts and borders of massifs of ancient lithosphere inthrust into geosynclinal provinces, 33—some other tectonic boundaries of different order.

related to the structures of the geosyncline frame, but not to the series which fill in the Koryak geosyncline proper.

The history of development of the territory over North-East Russia during the Mesozoic reflects the process of formation and growth of the continental crust in the northeastern margin of Asia. At the same time, it is largely a history of the opening up of major marginal seas in the Pacific surround which do not exist now but changed into the folded country of the Koryak Highland. It is reasonable to start considering the history of development of the entire territory from its most ancient part—the margin of the Mesozoic Paleo-Asian continent making up the hanging wing of the seismofocal zone which restricted the Koryak geosyncline.

MAIN TECTONIC FEATURES OF BASEMENT OF THE OKHOTSK-CHUKOTKA VOLCANIC BELT AND ITS PERIVOLCANIC AREA

The hanging wing of the Okhotsk-Chukotka paleoseismofocal zone is of a complex and heterogeneous composition (Fig. 35). This scheme partly repeats the tectonic schemes of different scales which were made for the territory of Northeast Russia at different times by other authors (Ye. K. Ustiyev, Yu. M. Puscharovsky, K. Ya. Springis, S. M. Tilman, B. M. Chikov, V. F. Belyi, M. Ye. Gorodinsky, and others). The main peculiarities of the presented scheme are as follows: 1) representation of the site of outcrop on the Earth's surface of the ancient seismofocal zone (the latter is not in all schemes and if it is, the disposition of it is not unambiguous; we will dwell on this question below); 2) division of intrusions into two morphologic groups, one of which is represented by large fields of irregular outlines which may occasionally contain, besides Mesozoic, pre-Mesozoic intrusive formations related to the foundation of the volcanic belt, and the other made up of isometric intrusions which are often circular in their planar sections and accompanied by ring faults, occurring mainly over the `perivolcanic' area behind the volcanic belt; 3) in the tectonic zonation, the zones of one identifies paleocontinental slopes of different ages which developed on the margins of the blocks of the ancient continental lithosphere and separated these massifs from the geosynclinal basins. The existence of continental slopes as a special structural unit can explain the distribution and composition of specific sedimentary formations as well as some magmatic formations and ore deposits.

As seen from the tectonic scheme (Fig. 35), the hanging wing of the Okhotsk-Chukotka paleoseismofocal zone consisted of large fragments of the ancient continental lithosphere characterized by outcrops of deeply metamorphosed rocks of the Archean and Early Proterozoic ages (Okhotsk, Omolon and Eskimo massifs) and their separating geosynclinal areas of different ages. The Okhotsk and Omolon massifs can, in many of their signs, be considered to be detached parts of the Siberian Platform; therewith, the Okhotsk massif in its southern part (hidden beneath the Mesozoic effusions of the Ulya Depression) does not seem to have fully separated from the Siberian Platform (which was mentioned by Yu. A. Bilibin, even in the 1930s). The pre-Mesozoic history of development of the entire territory can be considered to be a disintegration of the Siberian Platform edge, with emergence of geosynclinal areas between the platform and its broken-off fragments, and their subsequent changes, fully or partially, into folded structures. The process was of intermittent character. The geosynclinal structures which formed near the edge of the Siberian Platform and folded during the Middle Paleozoic were disrupted and deformed again during the opening up of the Late Paleozoic Verkhoyansk geosynclinal system. The latter, in turn, became a folded area during the latter half of the Mesozoic,

In the western half of the Okhotsk-Chukotka volcanic belt, where the latter and its foundation have no land border with the Koryak geosynclinal area (which apparently is hidden there in the northern Okhotsk Sea), the interrelation of the Koryak system's geosynclinal structures and its northwestern border, which are of interest to us, are not traceable. So let us consider the development of the foundation in the volcanic belt and its `perivolcanic' area in only its eastern half, which is immediately adjacent to the part of the Koryak geosyncline exhibited on land; much information on the development of this area can be found in works by a number of authors, in particular, B. M.Chikov, R. B. Umitbayev, F. F. Veldjyaksov, G. A. Grinberg, S. D. Sokolov, A. D. Chekhov, and other researchers.

OLOI—ANYUI FOLDED SYSTEM IN WESTERN CHUKOTKA

In the author's opinion, the time of the geosynclinal folding system does not depend on the time of formation and duration of the geosyncline development. The substantiation for this principle can be found in Chapter VIII. It is also supposed that the time spans during which the geosyncline experienced the folding can differ from place to place, i.e. the closure time in the different parts of an initially single geosynclinal system can be different. Hence, in this work, when the ages of geosynclines and their basic features are characterized, the reader will find separate indications for the time of existence of the geosyncline and the time when the folding occurred in it; unless a special explanation is given, the time of existence of the geosyncline is meant.

We attach the name of the Okhotsk-Anyui system to that system which is geosynclinal in its origin and folded in its present state, located within Western Chukotka, in the basins of the Anadyr River, Great and Little Anyui Rivers and Oloi River. This system of the northwestern strike comprises here the foundation of the Okhotsk-Chukotka volcanic belt and of the `perivolcanic zone' adjacent to it on the north-west. In the south-west, the Oloi-Anyui folded system borders the Omolon massif of the ancient continental lithosphere; in the north-east it is limited by the Alyarmaut rise and its contacting structures which seem to relate to the margin of the Hyperborean Platform which is now hidden on the shelf of the Arctic Ocean. The Oloi-Anyui geosynclinal system existed in the first half of the Mesozoic, Late Paleozoic and, probably, even earlier, although the last statement could be substantiated only by indirect evidence. The width of the Oloi-Anyui folded system from south-west to north-east is now 530 km.

According to Shilo *et al.* (1973), the Oloi-Anyui geosyncline system is an eastern continuation to the Paleozoic-Early Mesozoic Alazeya-Oloi system shown in the place of the Kolyma middle massif which was previously shown in most tectonic schemes. Without sharing the opinion on the relation of the Alazeya Rise and its contacting the Selenyakh—Uyanda inter-river space to the geosynclinal area *(*Shilo *et al.*, 1973, Tilman *et al.*, 1977*)* the author, however, supposes it to be probable that there exists a most ancient continuation to the geosynclinal system in the southern portion of the Kolyma massif along the north-western edge of the Near-Kolyma Rise. From this viewpoint, the Kolyma massif is a heterogeneous formation consisting of both the blocks of the pre-Late Proterozoic continental lithosphere in the northern part and the folded structures of Early-Middle Paleozoic age. It seems unlikely that after-Middle Paleozoic geosynclinal formations should exist in this province.

Judging from the similarity between the deposits, and from the age and the strike of the folded structures, the North-Chukotka (Chaun-Palyavaam) Early Mesozoic folded province located to the north-east of the Chaun Bay, is a most probable continuation of the Oloi-Anyui system. However, it proves to be displaced with respect to the strike of the Oloi-Anyui system by about 250 km to the north (Figs. 35, 36). The absence of the appropriate changes in the strike of the folded structures and an abrupt line of the displacement coinciding with a group of submeridional large faults allows an admission that it was caused by the horizontal shear of the corresponding amplitude. Horizontal displacements of this type, in particular, along large shears, do not seem a rare exception in the rear of the Okhotsk-Chukotka seismofocal zone, which will be shown below.

The tectonics of the Oloi-Anyui geosyncline system were described in works by S. M. Tilman, A. Ya. Radzivill, Yu. M. Dovgal, V. V. Gulevich, B. F. Palymsky, M. Ye. Gorodinsky, K. B. Seslavinsky, A. I. Afitsky, A. D. Chekhov and other researchers. Rather different opinions were expressed on the actual structures, within the considered area; however, the investigations allow us to surmise on the existence there in the first half of the Mesozoic of several geosynclinal troughs of a predominant north-west strike which are now represented mainly by the folded zones with the same strike.

Geosynclinal zones

Identified among the folded geosynclinal zones from north to south are: North-Anyui, South-Anyui and Krichal (Fig. 36).

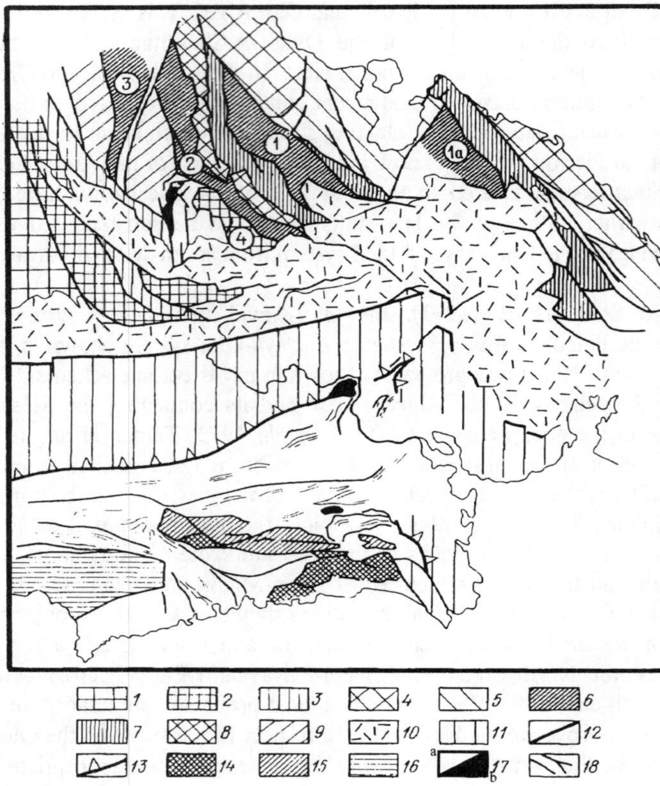

Fig. 36. Oloi-Anyui geosynclinal system in its interrelation with the Koryak geosyncline system. 1-9—Oloi-Anyui geosynclinal system; 1-5—its surrounding structures: 1—Omolon massif, 2—areas of its slope with outcrops of Paleozoic formations, 3—same, beneath cover of Mesozoic deposits, 4—margins of Hyperborean platform (Alyarmaut and Kuyul massifs) with outcrops of Paleozoic formations, 5—same, beneath cover of mainly Mesozoic deposits; 6-9—structures of the geosynclinal system: 6—troughs, 7—trough slopes (on non-geosyncline base), 8—Nutesyn (Nutesyn-Kameshkovka) Rise with outcrops of Paleozoic formations and volcanic covers of Upper Jurassic and Lower Cretaceous age—likely island arc in Triassic, 9—hidden beneath cover of Jurassic and Cretaceous sedimentary and volcanic deposits massifs with probably pre-Mesozoic foundation dividing geosyncline troughs. 10-16—Structures of surroundings of the Koryak geosyncline system: 10—volcanic covers of Okhotsk-Chukotka volcanic belt, 11—paleocontinental slope of Koryak geosynclinal basin (during the Jurassic and Neocomian), 12—upper restriction of this slope (frontal tectonic scarp on the paleocontinent's edge), 13—line of outcrop on the Mesozoic Earth's surface of surface of eduction out-thrusting (axis of Jurassic-Neocomian trench); 14-16—probable remains of outer island arc of Koryak paleobasins: 14—with recent outcrops of Paleozoic formations, 15—beneath cover of Mesozoic deposits (internal slopes of paleoisland arc), 16—beneath Cenozoic deposits. 17—Elements of inner structures of gosynclines: a - strike of fold structures, b - large ultrabasic massifs. 18—Some large faults. Encircled figures—structures of Oloi-Anyui system: 1—North-Anyui geosynclinal trough, 1a—its likely continuation to north-east (Chaun area), 2—South-Anyui geosynclinal trough, 3—Krichal geosynclinal trough, 4—Yarakvaam Rise.

The North-Anyui zone borders immediately upon the Alyarmaut Rise made up of relatively gently-dipping middle-Paleozoic sedimentary rocks and large massifs of granitoides. The zone, 100 km in width from north-east to south-west, is represented by Triassic terrigene deposits of all the three divisions, but mainly of the Upper Triassic, which is 6 km in thickness and deformed into folds with a north-eastern strike. The axes of the folds and strike of layers in their wings are not parallel to the border with the Alyarmaut Rise, but are cut by the latter at an acute angle which may testify to large displacements occurring along this border. The steepest and most complex folds are distributed along the northern margin of the zone; also found here were complex overthrust structures and nappes with the displacement directed both from south to north and from north to south. In contrast, the south-western part of the zone is characterized by a mild folding with a relatively gentle occurrence of the layers over wide areas. In the larger portion of the zone emerging onto the surface are mainly Upper Triassic deposits: the outcrops of the Middle and Lower Triassic deposits are concentrated in its northern half.

The South-Anyui zone which is to the south of the North-Anyui zone is a narrow band, 10 to 30 km in width, displaying a north-western strike and whose structure is made up of Upper Triassic, Lower, Middle and Upper Jurassic deposits, and those of the Neocomian, including the Hauterivian. These sediments, no doubt of considerable thickness, are much more strongly deformed than in the North-Anyui zone — they are crumpled into narrow, often compressed, isoclinal folds of a mainly north-western strike and constantly linked with numerous disrupting disturbances of the same direction; essentially, the rocks of the zone make up a great number of tectonic lenses and wedges whose initial interrelations are difficult, and occasionally impossible, to reconstruct. Owing to development of these complex linear structures, and the ubiquitous schist-forming process, the zone stands out from among its neighbouring areas (when viewed from the air, for instance). The zone, represented in its north-western portion by mainly Triassic deposits, also tapers towards the south-east (approaching the valley of Little Anyui); however a probable continuation of the zone to the north-west is made up by the Upper Jurassic sea deposits outcropping on both sides of Little Anyui and accompanied by a series of small hyperbasite outcrops in the basin of the River Yarkan.

The Krichal zone is to the west and south-west of the South-Anyui zone, largely beyond the scheme shown in Fig. 36 within which one can see only its south-eastern portion bent to the south. The strike of the zone in its main portion is steadily north-western. It is represented by the Upper Triassic and Jurassic deposits of all three divisions (but mostly of the Upper one) which are crumpled into steep folds with north-western strike.

In the south-eastern termination of the zone, in addition to sedimentary deposits, there are outcrops of metamorphosed basic rocks of the basement of the geosyncline (the basin of the River Bannaya) which are usually regarded as deeply altered effusions.

Among the other areas of clearly geosynclinal origin in the territory under consideration, one should point out the valley of the River Great Anyui in the vicinity of the River Aluchin mouth to the south of the South-Anyui geosynclinal zone. Opposite to the River Aluchin on the right-hand bank of the Great Anyui, there is a large bank of bedrock exposure with outcrops of faunally characterized Lower Permian deposits. They occur in the form of a gentle monocline inclined towards the north-east at an angle of 15-20° and are represented by interbedded aleurolites and argillites, sandstones, tufogene sandstones of varying granularity, occasionally small-pebble conglomerates and bluish banded jasper-like schists. These are the most ancient faunally characterized amd probably geosynclinal deposits within the Oloi-Anyui geosynclinal system. The reason for this conclusion is that they are directly overlaid by rocks of volcanic flows of spilite and keratophyre composition at the foot of the exposure, and also their position amongst the field of much more deformed and changed terrigene deposits which alternate with exposures of metamorphosed basic rocks.

Similar to the above described and, probably, analogous to them in age and origin are terrigene and basic formations which are widespread on the right-hand bank of the Great Anyui, 120 km to the east, in the basins of its right-hand tributaries — the rivers Bystryanka and Nizhny Megerenvei.

Southern margin of the Oloi-Anyui geosynclinal system and Yarakvaam Block

As seen from the above, the geosynclinal zones proper comprise only a part of the area occupied by the Oloi-Anyui geosynclinal system. As to the structures which make up their fringes, the authors differ much more. Two opinions were offered on the nature of the area between the South-Anyui zone and the Omolon massif (if one considers the Berezovaya Fault zone passing over the left-hand territory of the River Oloi to be its border). In the opinion of Radzivill, Dovgal, Palymsky, Gorodinsky, Seslavinsky and others, this area as a whole is a lowered edge of the Omolon massif which experienced tectonic deformations during the Mesozoic. According to another viewpoint, suggested by Tilman, much of this area also makes up a separate, and also partly submerged, rigid massif of a relatively ancient, supposedly Paleozoic, consolidation which was called Yablon. This massif is separated from the Omolon one by the zone of the Yeropol rise which, according to Tilman, corresponds to the Paleozoic eugeosyncline.

In the opinion of the author of this book, the Yeropol Rise corresponds to the paleocontinental slope of the Omolon massif at its border with the adjacent geosyncline. Within the rise there are no formations which would speak in favour of its belonging to the eugeosyncline: in particular, there are no hyperbasites, or spilites, or similar rocks but there are widespread terrigene Devonian and Carboniferous deposits of considerable thickness, including sandstones with a graded lamination, grits, cherty schists, and occasionally organic limestones, i.e. a complex of rocks typical of continental slopes. Some fields with signs of a rigid basement to the north of the Yeropol Rise should in this case be related to the lower part of the continental slope which is not hidden beneath Jurassic and younger sediments. The existence itself of the paleocontinental slope between the main part of the Omolon massif to the south-west of the Berezovaya Fault, repeatedly exposed above sea level in the Devonian and Carboniferous, and the geosyncline area seems to be beyond doubt. If the north-east border of the Omolon massif together with its accompanying continental slope bends when approaching the Okhotsk-Chukotka volcanic belt, as can be judged from the tectonic boundaries of the Yeropol Rise, then the width of this zone of ancient slope is about 120-150 km, i.e. corresponds to a standard width of continental slopes (see Figs 35, 36). However, even with this supposition, according to which the paleocontinental slope of the Omolon massif is brought near to the southernmost of the obvious geosynclinal structures—the South-Anyui geosynclinal zone, one is still challenged with a need to explain the extent of its advance to the north (the same may be said of the edge of the Omolon massif according to Gorodinsky, or to the Yablon massif according to Tilman); the explanation of this phenomenon is offered below.

The Yarakvaam Block may represent an especially far to the north displaced part of the Omolon palaeocontinental slope, of its deformed portion considered above. It is a positive structure which is elongated over the north-western direction and has a width of up to 30 km and a length of over 150 km (Fig. 36). The eastern and western halves of the structure display outcrops of acid and medium lavas, tuffs and tuff breccias which are characterized by poorly preserved fauna (*Productids*, etc.) . These were described by Dovgal, Radzivill and other researchers, who proceeded from the probable age and lithologic analogies, to conclude that the Omolon massif belongs to the Kedon series of the Middle—Upper Devonian. The Devonian extrusions have signs of partially sub-aerial origin (the evidence cited by Tilman on the deep-sea spilites occurring in the block is most likely related to the Ust-Orlovka field adjacent to the block on the west and considered in the above description of the geosynclinal areas). In the western portion of the block, Lower Carboniferous volcanogenic deposits containing the remains of shallow-sea fauna and sub-aerial flora are known; there are also (at Belyi Creek, according to Dovgal) outcrops of mica-quartz, garnet-biotite-quartz and epidote-actinolite metamorphic schists, probably belonging to the metamorphic basement of the block. The eastern half of the block is a fragment of a gently-sloping dome with extant slopes facing west and south-west, covered by Devonian and Carboniferous volcanic formations, shallow-water Lower Permian sediments and also shallow-water volcanogenic-sedimentary deposits of the Triassic. Both of the latter form series of only some hundreds of metres.

The western half of the block, unlike the eastern one, is of barchisyncline character. The provisionally Devonian volcanites are overlapped here by Upper Triassic deposits with the remains of shallow-water flora and fauna, and over them are gently dipping Jurassic shallow-water deposits with a thickness up to 1.5 km occupying a considerable part of the area. In some places, Jurassic deposits are overlain with sub-aerial Cretaceous effusive covers of acid and intermediate composition.

If one agrees with the relation of the Devonian volcanic rocks to the Kedon series, which is typical of the inner part of the Omolon massif, then a supposition is possible that the Yarakvaam Block is indeed a part of the Omolon massif and was later displaced and tectonically introduced into the geosyncline province (with a likely amplitude of horizontal displacement of up to 150 km). As was noted above, such extensive shifts during the folding transformation of the Oloi-Anyui system are quite possible.

Structures dividing geosynclinal zones

An essential difference between the treatment of the tectonic structure of the Oloi-Anyui province supported by the author and the one usually used consists in the comprehension of the nature of those structures that divide the above-described geosynclinal zones. These structures dividing and restricting the geosynclinal zones are largely represented in the recent erosional surface by the Upper Jurassic, Neocomian and Aptian-Albian volcanic deposits and their accompanying sediments, whereas the neighbouring geosynclinal folded zones mainly accommodate the outcrops of Triassic and Jurassic sediments. The presence of outcrops of young deposits among more ancient ones is often considered to be a sign of a negative tectonic structure of synclinal or graben type. For example, between the North-Anyui and South-Anyui zones, there can be identified the Yarkan-Nutesyn (or Nutesyn) `depression', whose north-western continuation (on the right-hand region of the River Small Anyui) is called the Kameshkov depression; between the north-western part of the South-Anyui and Krichal zones, there is the Orlovka depression, and to the south of the Krichal zone there is the Khetachan (Khetachan-Kurya) depression (Geology of the USSR, 1970). Obviously, it was thought that these depressions emerged on the geosynclinal folding basement similar to the neighbouring geosynclinal zones. However, on the southern flank of the Khetachan-Topolevka structure and in the Yarkan-Nutesyn structure, one can find outcrops of Paleozoic deposits, including organic limestones (conventionally dated in the former case as the Middle—Late Devonian and in the latter case, Early Carboniferous). Within the geosynclinal zones, such outcrops were not found, but they are present in the Yarakvaam Block.

Let us dwell at greater length on the *Yarkan-Nutesyn structure*. It is a band between the North-Anyui and South-Anyui geosynclinal zones of a width of 10-15 km and length of over 150 km. The band comprises rather gently deformed Upper Jurassic, Neocomian and Aptian-Albian mainly sub-aerial volcanic rocks of intermediate and basic composition accompanied by shallow-water sediments with fauna and flora (the Nutesyn suite). The base of these deposits is exposed only in the River Second Topolyevka, where there are the above-mentioned provisionally Carboniferous deposits with sediments of the Upper Triassic unconformably overlying them. The latter are of insignificant thickness which is much less than in the neighbouring geosynclinal zones, and it is this that allowed the emergence of the Paleozoic rock outcrops on the recent surface. Smaller thicknesses, stratigraphic discontinuities, and sub-aerial conditions for accumulation of the Upper Jurassic—Neocomian deposits clearly indicate that during the Late Jurassic and Neocomian, the Yarkan-Nutesyn structure was elevated relative to the neighbouring South-Anyui zone, where similar-age deposits can be seen in the rather deep terrigene sea facies. The same is true of its position both with respect to the South-Anyui and North-Anyui zone during the Late Triassic and earlier, since the thickness of the Upper Triassic sediments in the Yarkan-Nutesyn structure proves much less than in these geosynclinal zones, while the Middle and Lower Triassic sediments which are widely distributed in the North-Anyui zone, in this structure did not sediment at all, or else they were eroded prior to the Late Triassic. Consequently, during the Triassic, the Yarkan-Nutesyn zone was a stable rise, occasionally elevating above sea level. During the Jurassic and Neocomian, it preserved its position with respect to the South-Anyui trough although the absence of the appropriate deposits from the North-Anyui zone seems to testify that by that time it had already undergone a folding and was most likely a new mountain structure. The Yarkan-Nutesyn structure was thus a kind of step at the foot of the highland above the trough of the South-Anyui zone. Only after folding, which involved all the South-Anyui trough at the end of the Neocomian, could the Nutesyn structure show itself for some time (probably, before the end of the Albian) as an intermountain area; later on it became a part of the general mountain structure.

Thus, during the geosynclinal stage of development of the territory, the Yarkan-Nutesyn structure was a constant rise between the North-Anyui and South-Anyui troughs; the preservation of rather young covers at its limits, if compared with the surrounding areas, is the consequence of the stability of its level, which also persisted at the stage of the folding: when the relief of the neighbouring troughs underwent an inversion and turned into a highland, its rising was slower and it was less subjected to erosion. At present, the Yarkan-Nutesyn structure occupies once again a relatively elevated position making the main divide between the basins of the Great and Small Anyui rivers.

The conclusion expounded is in agreement with the logical necessity of the existence of a positive structure, which divided the North-Anyui and South-Anyui troughs which were thus different in their structural development and history. This conclusion also accounts for an obvious fact of weaker deformations of the Upper Jurassic and Neocomian deposits within the Yarkan-Nutesyn structure as compared with South-Anyui trough: the former seems to have a more rigid foundation than in the geosynclinal zones. All this makes it reasonable to think that the Yarkan-Nutesyn structure was, at least during the Triassic, a stable elevation of an island arc type between the North-Anyui and South-Anyui geosynclinal troughs. The lower age limit of its emergence can be found from the outcrops of conventionally Lower Carboniferous deposits along the River Second Topolevka and is, thus, the Late Paleozoic. The Yarkan-Nutesyn zone was the probable crest of the arc, but there was probably also its slope within the southern part of the North-Anyui zone which is characterized, as was mentioned, by comparatively weak dislocations of the Upper Triassic deposits. The total width of the island arc foundation was in that case 50 to 60 km. Proceeding from comparisons admitted for the

Paleozoic outcrops of the Yarkan-Nutesyn rise, as well as from its position, one could believe that the island arc's foundation was torn away from its northern Alyarmaut Rise and adjacent structures during the development of the Oloi-Anyui geosyncline system.

It can be envisaged that some other above-mentioned structures, often treated as structural depressions, belong to the rises with non-geosynclinal rigid basement. In addition to their similarity to the Yarkan-Nutesyn rise in their recent geological structure, this conclusion is arrived at from their position between individual geosynclinal troughs. Above, it was mentioned that Paleozoic outcrops are present within the Khetachan structure. There seems to be another reason for presentation of these structures on geological maps as depressions: within them, including the Yarkan-Nutesyn structure, there are distributed the fields of volcanic rocks of the Late Jurassic — Neocomian — Aptian-Albian age, while there are very few or none of them in the folded zones which correspond to obvious geosynclinal troughs. One cannot be sure enough whether they existed there and were fully eroded away or if they had never existed at all. However, clear tectonic boundaries of most areas with volcanic covers, which in many cases are conformable with the adjacent folded structures, indicate that they are structures of a block type with which volcanic covers seem to be justly associated. As follows from what was said about the Yarkan-Nutesyn structure, the foundation of these blocks probably comprises the detached parts of the ancient consolidated lithosphere from the surround of the geosynclinal system. The connection of predominant volcanic out-flows with such detached blocks seems to be probable in view of the undoubted fact of spatial coincidence of the widely occurring different-age volcanism (Devonian and Cretaceous) with the large massifs of ancient continental lithosphere at Okhotsk and Omolon. We will not dwell here on the cause of this, but only note that the supposition explains a finger-like mosaic arrangement of volcanic fields within the Oloi-Anyui folded system which seemed to be quite irregular at first glance: it depends on the existence amongst the geosynclinal area of some blocks of rigid lithosphere accompanying particular volcanic activity.

Reconstruction of the geosynclinal system pattern and age of folding

Independent of treatment of the structures in the Oloi-Anyui system, there is no doubt that they underwent large displacements and deformations. From the above viewpoint, the reconstruction of the initial structures of the geosynclinal province can be conceived in the following way. The axis of the main geosynclinal trough most likely corresponded to the South-Anyui zone. This zone of the most intense extension in the past, seems to have been most susceptible to folding deformation and has been turned into a narrow, most deformed structure which in places is similar to a suture seam. In spite of this feature of its recent appearance, residual geosynclinal basins were preserved within it for an extended time (at least up to and including the Hauterivian). To judge from the initial stage and early existence of this zone, one can use data on geosynclinal formations of its immediately close Ust-Orlovka geosynclinal field where this should be considered pre-Permian.

The fact that there are no pre-Permian fauna in the other troughs of the Oloi-Anyui system cannot disprove this conclusion, as one should take into account that early sediments in the opening geosynclines are of smaller volumes than are the later ones; besides, they are largely deep-water and, thus, seldom contain faunal remains. The intense deformations that were undergone by these geosyncline structures make it still harder to detect them. The conclusion about the comparatively ancient age of the initial geosynclinal troughs also fits the distribution of the Devonian deposits of all three divisions and of Lower Carboniferous terrigene deposits on the slopes of the Omolon massif (with older deposits on its farthest western slope).

Predominantly acid sub-aerial, terrestrial, volcanites on the Omolon massif (as one can judge from an analogy with the Mesozoic Okhotsk-Chukotka volcanic belt) point to the geosyncline that was opening up nearby (the latter opinion, as well as the concept of the existence of the Paleozoic geosyncline in the Oloi area, has long been suggested by S. M. Tilman).

The structure of the northern part of the Oloi-Anyui geosynclinal province has already been treated above. At some stage following the Early Carboniferous, most likely at the end of the Paleozoic, the Yarkan-Nutesyn island arc broke away from its northern border forming a marginal sea to the rear which corresponds now to the North-Anyui zone. It is hard to say whether such structures had existed anywhere on the southern margin of the central geosynclinal trough adjacent to the Omolon massif.

The opinion expounded here on the correspondence between the central trough of the geosynclinal system and the South-Anyui zone differs from the views presented in this connection by both Tilman and other researchers. According to Tilman, such a trough was located at theYeropol Rise and was made up of mainly Devonian deposits. Reasons were given above which allowed us to relate this rise to the slope of the Omolon massif. According to Dovgal, Radzivill, Gorodinsky, Seslavinsky and some other researchers, the main geosynclinal trough was the North-Anyui one (called just Anyui). These authors take into consideration a greater width of the North-Anyui zone compared with the South-Anyui one and a continuous stratigraphic section of all Triassic divisions in it. However, it was shown above that the time of the initial stage of the South-Anyui trough with its environs should be considered to be older than Triassic. Structural interrelations in the South-Anyui zone are more complex, as a result of much more intense tectonic deformations than in the North-Anyui zone. The latter fact testifies to the much greater plasticity of the geosynclinal basement in the South-Anyui zone.

Seslavinsky (1970) defended the opinion that the South-Anyui trough emerged as late as the Jurassic. Such a view is disproved by the remains of Late Triassic fauna found in the South-Anyui zone. However, Seslavinsky's evidence on overlaying by the Upper Jurassic sediments of the spilite-like rocks and ultrabasite protrusions near the northern margin of the South-Anyui trough (in the basins of the rivers Glubokaya and Yarkan) show that this part of the trough did indeed form during the Late Jurassic. This age of the geosyncline basement in front of the island arc (Yarkan-Nutesyn Rise) seems to be quite real, because the presence of an island arc testifies to the existence of a trench before its front and, if the process was proceeding eductionally, out-thrust of the geosynclinal floor would take place in it. Appropriately, the youngest geosynclinal foundation should be adjacent to the trench and the foot of the island arc (Chapter III; many more details of the eduction opening of the geosyncline can be observed in the Koryak district).

Taking the South-Anyui zone as the central geosynclinal trough, it is difficult to assume that there is no continuation of it to the west. In all probability, such continuation should extend along the Omolon massif or, at least, along its north-eastern margin (at a distance corresponding to the width of the hidden foundation of the continental slope). In composition of sediments and character of their deformation, the Krichal geosynclinal zone could be such a continuation. However, this explanation demands a large-scale right-side shift displacement (of up to 200 km) of the entire geosynclinal zone along the Nenkan—Great Anyui group of faults (this is a large group of rectilinear closely packed faults with north-western strike stretching from the River Nenkan reach to the River Bannaya mouth and the River Great Anyui valley). A supposition of such a shift is so far hypothetical; however, such assumptions have to be made repeatedly in analyzing the structure of the Oloi-Anyui geosynclinal province.

To the end of the Late Triassic, at any rate in the Carnian, all the geosynclinal troughs—the North-Anyui, South-Anyui and Krichal—were still areas of sedimentation. The earliest folding dislocations of the end of the Triassic or at the beginning of the Jurassic embraced the entire

North-Anyui trough where no sea sediments had accumulated since the end of the Triassic. (Some authors consider that the Jurassic and Neocomian deposits in the basin of the Rivers Rauchua and Pogyden east of the Alyarmaut Rise offer evidence for continuation of sedimentation after the Triassic in the north part of the North-Anyui zone; however, the Valanginian and Upper Jurassic deposits as well as their underlying Upper Triassic ones are gently dipping over the larger part of this area, which relates it to the submerged part of the Alyarmaut Rise, but not of the North-Anyui zone).

The dislocations of the same Late Triassic-Jurassic age also developed in the above-mentioned, probably displaced, continuation of the North-Anyui trough—the North-Chukotka, or Chaun, folding zone. The position of the North-Anyui geosynclinal trough between the Yarkan-Nutesyn island arc and the northern fringe of the geosynclinal province, i.e. the position of a marginal sea, allows one to think of it as a folding transformation with the continued opening up of the central South-Anyui trough, arising from compression which may have developed in the hanging wing of their dividing, probably paleoseismofocal zone. Its corresponding eduction surface must have emerged onto the surface sea floor before the front of the Nutesyn island arc to make a border of the latter with the South-Anyui trough. Proof for the existence of the paleoseismofocal zone can be seen in the above-cited evidence of the continued opening up of the South-Anyui trough, in spite of the folding in the North-Anyui zone.

The Late Triassic or Early Jurassic folding in the North-Anyui zone does not testify, according to what was said, to distribution of the folding over the entire Oloi-Anyui geosynclinal area, but marks the beginning of its tectonic reconstruction. The later geological history comprised a successive involvement of geosynclinal troughs and their parts into folding deformations to turn them into mountain structures. At the end of the Jurassic, the folding process extended into the Krichal geosynclinal zone; during the Neocomian, the sediments probably did not accumulate within it, although the northern part of the zone is hidden under the Quaternary deposits. However, the shallow sea Neocomian deposits are rather common in its adjacent, supposedly more rigid, Kharchan structure on the south and the Orlovka one to the north, as well as to the south on the slope of the Omolon massif in the basins of the rivers Penjenka, Oloichik, Oloi and others. These structures of the geosynclinal surround and its detached parts, after the folding of the geosynclinal troughs, turned into depressions among the new mountains which, in most cases however, did not have their own descending movements. Crumpling of the Krichal geosynclinal zone could have been due to the northward displacement of the Omolon massif against whose north-eastern edge it was positioned.

According to the fauna of the South-Anyui geosynclinal trough, sedimentation continued into the Berriasian, Valanginian and Hauterivian. The difference in the time of the closure of the Krichal and South-Anyui zones may indicate that their separation, if they had been initially a single zone, took place prior to the Neocomian. According to Seslavinsky (1970), to the northern part of the South-Anyui trough the Neocomian deposits occur on the Upper Jurassic ones with slight unconformity or with a gradual transition. It is not ruled out that the Neocomian sediments tending to the northern boundary of the South-Anyui trough, that are coarser and more shallow-water compared with the Upper Jurassic sediments, filled the area of the trench which had existed in front of the Yarkan-Nutesyn island arc. As follows from the above, the folding in the South-Anyui zone is post-Hauterivian. At the same time, a number of its intrusions which had been injected into the previously formed folding structure (intrusions of the Vulkannaya, Blokhin and other mountains), according to determination of an absolute age, relate to the end of the Early Cretaceous or beginning of the Late Cretaceous (about 100 mybp). Thus, the folding in the South-Anyui trough took place obviously in the Barremian—Aptian.

We have pointed out the complexity of the fold and block structures in the Oloi-Anyui geosynclinal province, as well as the possibility of their different treatment. However, with all reconstructions, irrespective of the nature of individual structures, a hard conclusion is that their general strike is clearly north-western and the geosynclinal conditions predominated there up to the end of the Triassic.

INTHRUST MASSIFS

We have dwelt upon the development of the Oloi-Anyui province for reasons of necessity to clearly differentiate between the structure of the Oloi-Anyui and Koryak geosynclinal systems. Although in the Oloi-Anyui system the signs of geosynclinal development can be seen clearly enough, the basic evidence supporting the eduction origin of the marginal geosynclines in the North-East of Russia can be obtained in the Koryak geosyncline system which is younger and easier to access for historical analysis, in spite of its very complex internal structure. However, before describing the latter, let us consider peculiar forms of large tectonic displacements which seem to be characteristic of the hanging wings in many seismofocal zones, including the north-western wing of the Okhotsk-Chukotka paleoseismofocal zone, which includes both the Koryak and Oloi-Anyui geosynclinal areas.

Arch-shaped structures of the hanging wing of the Okhotsk-Chukotka paleoseismofocal zone. The outlines of the Okhotsk and Omolon massifs (of ancient continental lithosphere) share much in common: they have a base which is broad in plan, facing south-east towards the line of outcrop of the paleoseismofocal zone, and an arch-shaped (parabolic or similar in form) opposite boundary facing inside the continent (see Fig. 35). The arch-shape borders of the massifs are clearly bent around by folding structures of adjacent geosynclinal areas. Especially clear is the enveloping of north-facing borders of the Okhotsk massif with folding structures of the geosynclinal Verkhoyansk Late Paleozoic sedimentary complex. Adjoining the Okhotsk massif on the north-west, the folding structures of the north-eastern strike bent near the northern border of the massif in the basin of the River Indigirka, become latitudinal in strike; farther to the east and opposite the north-eastern border of the massif, they strike from south to west. The age of folding in the Verkhoyansk geosynclinal complex was found by many researchers to be of the Late Jurassic—Early Cretaceous; this was the time of great activity within the Okhotsk-Chukotka paleoseismofocal zone (for more detail, see below). Obviously, in the rear of this zone, tectonic processes forced the inthrust of the rigid Okhotsk massif with its ancient Archean foundation into the Verkhoyansk geosynclinal province (which was opening from the Late Paleozoic to the end of the Jurassic—*Geology of USSR*, 1970; and others). The same may be true of the equally ancient Omolon massif with the difference that its folded surroundings, which are partly more heterogeneous in age and are hidden beneath the deposits of the near—Kolyma lowlands, are traceable over disconnected sections.

The Okhotsk and Omolon massifs are thus a sort of rigid die-stamp inthrusting into more plastic geosynclinal basements. The existence of inthrust massifs seems to be typical of rear parts of the above-seismofocal areas, as are the convex shapes of their borders facing inside the continent. From this point of view, of special interest is the Aluchin structure in the basin of the Great Anyui River, smaller in size and internally heterogeneous but also of a regular arch-shape. This structure is highly conspicuous on geological maps of any scale owing to the cutting on its convex western border, sharply pronounced due to its associated large Aluchin ultrabasic massif, of the strike of the folded formations that are distributed westwards. (Herein we use the name Aluchin for a larger structure than that which is presented on many tectonic schemes with the name of the Aluchin block and corresponds to the district where the Aluchin ultrabasic massif is disposed; the latter is a part of the frontal zone in the Aluchin structure according to our description).

The Aluchin structure (Fig. 37) occupies a large part of the basin of the upper reaches of the Great Anyui River and is about 120 km in length. In the east, its widening base is hidden under the volcanic Cretaceous covers. The structure has a complex interior with heterogeneous surroundings but its borders are clearly visible owing to changes of structure and geological complexity. The northwestern border of the Aluchin structure is represented by a series of dislocations of a complex shape in plan and convex towards the northwest which, according to the pattern they form, are mainly of southwest direction in the dip of their planes. Occasionally, in this frontal part of the structure, one can observe series of plate-like intrusive bodies of variegated composition (from plagiogranites to gabbro-diorites) which filled the southeast-dipping tectonic fractures with dip angles of 60 to 45° (in the bank outcrops of the Great Anyui River opposite the Aluchin River mouth). In the north of the frontal zone, in the interfluve of the rivers Monni and Orlovka, there seem to occur gentle-dipping overthrust slabs with tectonic erosion windows in them. In the northeast and southwest, the tectonic borders of the Aluchin structure become more rectilinear and show signs of large slip displacements. The displacement pattern is most conspicuous on the southwestern border of the structure. One can see here a series of closely packed rectilinear faults of northwestern strike with tectonic wedges among them made up of volcanites of a Kedon type of the Middle—Upper Devonian complex, and a wedge of ultrabasites displaced to the southeast of the main Aluchin peridotite massif. The disposition of the wedges indicates left-hand shear displacements of blocks with a likely amplitude of dozens of kilometres. On the northeastern border of the Aluchin arch-shaped structure its correlation with the elements of the South-Anyui zone indicate large horizontal right-hand slips.

Along with the Aluchin ultrabasic massif, the Aluchin structure comprises, near its western border, the above-mentioned Yarakvaam Block, 40 by 10 square kilometres in extent, and, adjacent to its north-eastern part, the Ust-Orlovka field of distribution of the Permian and pre-Permian geosynclinal formations (opposite the north border of the Aluchin ultrabasic massif). To the northwest of the northwestern part of the Yarakvaam Block and Ust-Orlovka field, there are fields of complexly alternated ultrabasic and gabbro rocks which probably should be regarded as a series of melange blocks and flake blocks of dynamometamorphosed basic and ultrabasic rocks. They are often considered to be Devonian or Permian alternating volcanites, but in some part, indubitably, they are metamorphic rocks of basic composition.

The central part of the Aluchin arch-shaped structure is occupied by the Ainakhkurgen depression filled with mainly Aptian-Albian and partially with Barremian continental deposits. As seen from Fig. 37, the shape of this depression is roughly similar to the outlines of the Aluchin structure border as a whole; this shape is also repeated in the strikes of some tectonic Aluchin slopes and folded structures in the inner parts of the basin.

A considerable part of Ainakhkurgen depression basement seems to belong to a rigid massif comprising the southern part of the Aluchin structure (a part of the Yablon massif according to Tilman). It outcrops onto the surface from under the youngest Mesozoic deposits of the Ainakhkurgen depression on its east side, between it and the Aluchin peridotitic massif, and is represented there by a weakly deformed cover of Jurassic and Neocomian deposits. On the left-bank region of the Great Anyui River, below the River Aluchin mouth, outcropping from beneath this cover there are Paleozoic granitoides of an absolute age of 220 my.

The direction of displacements of the Aluchin structure on the whole seems to be obvious enough from the probable directions of the slips on its sides which are seen from the structural

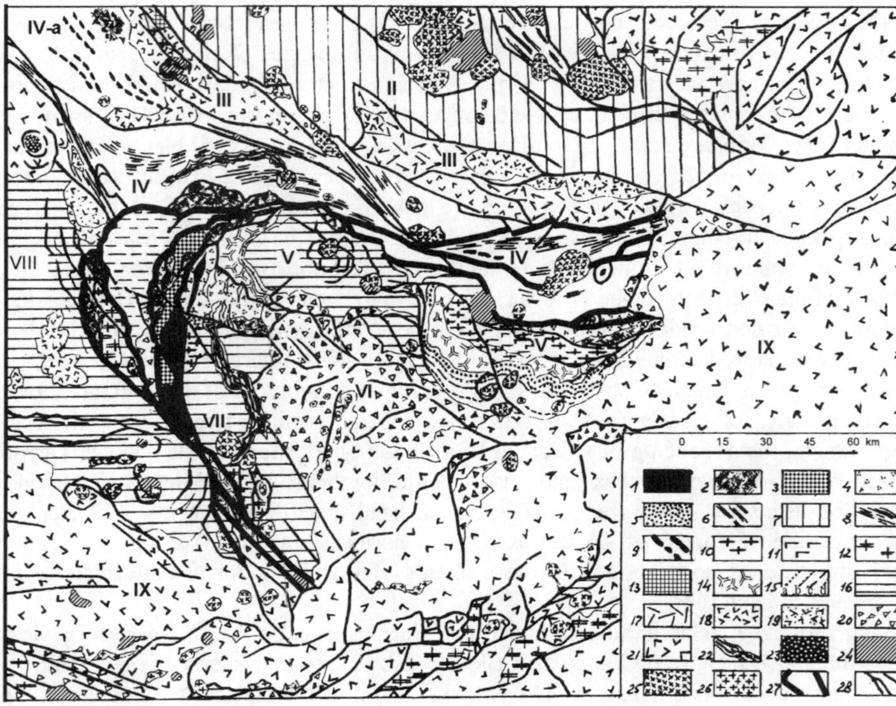

Fig. 37. Aluchin arch-shaped structure. 1—ultrabasites, 2—micromelange, 3—localities with developed large-block melange and gabbroids, 4—essentially dynamometamorphized rocks of supposedly Middle—Upper Paleozoic, 5—Late Permian geosynclinal deposits, 6—Triassic geosynclinal deposits in North-Anyui trough, 7—southern slope of North-Anyui trough (northeastern slope of Nutesyn Rise), 8—Upper Triassic, Jurassic, and Neocomian geosynclinal deposits of South-Anyui trough, 9—same Jurassic in northern part of trough, 10—pre-Mesozoic granitoids, 11—pre-Mesozoic gabbroids, 12—fields of granitoids and partly gabbroids, possibly, including pre-Mesozoic varieties, 13—acid and intermediate volcanites of Middle—Upper Devonian (analogues of Kedon series of Omolon massif), 14—Carboniferous deposits, partly coal-bearing, 15—subhorizontal and gentle-dipping deposits of Lower Permian (a), Upper Triassic (b), Lower and Middle Jurassic (c), and Upper Jurassic (d) ages on Yarakvaam Rise, 16—subhorizontal Jurassic and Neocomian deposits supposedly testifying to rigid foundation under them (a—slope of ancient Omolon Massifs with foundation of platform type, b—fold buildings of Paleozoic age), 17—acid, partly subaerial, volcanic covers of Late Jurassic age, 18—intermediate and basic essentially subaerial volcanic covers of mainly Neocomian age, 19—volcanogenic-sedimentary subaerial deposits of the late Neocomian, Aptian, and Albian ages, 20—clastic, mainly subaerial deposits of the Aptian and Albian, 21—subaeral volcanic covers of basic and intermediate, partly acid composition of Albian-Cenomanian age, 22—Upper Quaternary basalts; 23-27—intrusive rocks mainly in tectono-plutonic structures of central type: 23—basalts, 24—gabbroids, 25—granodiorites, 26—granites, 27—granosyenites; 28—tectonic boundaries of different kind (faults, horizontal slips, overthrusts). Figures: I—North-Anyui geosynclinal trough; II—southwestern slope of North-Anyui trough; III—Nutesyn (Nutesyn-Kameshkovka) Rise; IV—South-Anyui geosynclinal trough (IVa—its northern part with post-Triassic age of base sediments); V—Yarakvaam Rise; VI—Ainakhkurgen depression; VII—blocks from slope of Omolon massif inside Aluchin arch-shaped structure; VIII—same, beyond Aluchin structure; IX—Okhotsk-Chukotka volcanic belt.

features, and overthrust character of its northwestern frontal boundary. So the Aluchin structure, similar to the Okhotsk and Omolon massifs, proves to be an inthrust massif displaced to the inside of the continent but, unlike those others, of a heterogeneous composition. As mentioned above, a likely amplitude of the horizontal displacement of the Yarakvaam Rise may amount to 100-150 km. This conclusion does not contradict the pattern of the structures surrounding the Aluchin heterogeneous arch-shaped structure; an inthrust of the Aluchin massif accounts for the cause and character of displacement of the Yarakvaam Block with its facies of Devonian volcanites resembling those of the Omolon massif into the field of the geosynclinal deposits.

Displacement of the Aluchin structure had led to the overthrust and bulging up of its front edge which resulted there in the deepest erosion level within the whole Oloi-Anyui province (Aluchin peridotite massif and Ust-Orlovka field of the Paleozoic geosynclinal formations). At the end of the Early Cretaceous, owing to this bulging up of the northwestern part of the Aluchin structure, there arose a kind of a mountainous barrier whose destruction products were removed into the Ainakhkurgen inland basin which had formed in the middle of the Aluchin structure. The Aptian—Albian deposits which make up the Ainakhkurgen field contain a large number of pebbles of those rocks which outcrop on the western edge of the Aluchin structure, including ultrabasites and various intrusive rocks making up the above-mentioned fractured intrusive bodies in the northwestern frontal section of the structure (these have been pointed out by many researchers and confirmed by the author's own observations). The uplift of the northwestern part of the Aluchin structure was thus continual during the sedimentation in the Ainakhkurgen depression, and the intrusive bodies which injected during formation of the Aluchin structure were quickly brought up to the surface and subjected to erosion.

From the cited evidence, the time when the Aluchin heterogeneous arch-shaped massif began its displacements and, consequently, the time of its formation could be determined very exactly to be the Barremian—Aptian, since in the southeastern part of the South-Anyui trough, which is now adjacent to the northeastern border of the Aluchin structure, a residual geosynclinal basin still existed there during the Hauterivian, i.e. there were no signs of compression caused by a displacement of the Aluchin structure. Thus, the folded formations themselves in the South-Anyui basin were largely due to displacement of the Aluchin massif.

A peculiar feature of the Aluchin massif is that there are two Late Quaternary volcanic centres near its borders. Quaternary volcanism is a very rare phenomenon within the Okhotsk-Chukotka Mesozoic volcanic belt; the Late Quaternary outflows on the borders of the Aluchin structure are the largest known. The volcanic centre near the northeast border of the Aluchin structure is on its external side (the Anyui Volcano, or `the Volcano of the Young', within the round granodiorite massif of Mt. Vulkannaya). The youngest basalt flow which erupted from here (of an age of several hundreds of years, according to Ye. K. Ustiyev) filled the lower-lying valley of the River Monni for over 40 km. Another centre is near the southwest border of the structure from its internal side. A somewhat older basalt flow which erupted here filled the River Aluchin valley for a distance of 55 km. It is difficult to offer a definite tectonic treatment of this phenomenon: in all probability, it testifies to considerable depth of the tectonic boundaries of the Aluchin structure.

One can suggest that the Aluchin out-thrust massif is an internal part of a larger structure of a similar shape which is like the Okhotsk and Omolon massifs in size. The above mentioned zone of the Nenkan-Anyui faults, along which there might occur a large slip of the South-Anyui and Krichal geosynclinal zones, could be considered the southwest boundary of this (Great Anyui) structure. In the northwest part this structure may extend over the Alyarmaut Rise. However, unlike the Aluchin structure which is quite pronounced, it can be proposed only as a suggestion (Figs. 36, 38).

Some features of distribution of arch-shaped inthrust structures within the east margins of Asia. An instance of the Aluchin structure with its heterogeneous internal composition shows that the emergence of arch-shaped inthrust massifs in the rear of the seismofocal zones is not merely a result of the form of individual large blocks (as might be supposed for the Omolon and Okhotsk massifs) but is regularly caused by tectonic forces. In this connection, it would be appropriate to note that large arch-shaped tectonic structures with a convex sirface facing inside the continent are common over the entire Pacific margin of Asia to the south of the Okhotsk-Chukotka Mesozoic volcanic belt (accompanied with trenches, i.e. of an eduction nature from our point of view). Their sizes exceed those of the Okhotsk and Omolon massifs, let alone the Aluchin massif: each of them has extensive outcrops of the ancient platform foundation on the surface, although the composition on the whole is also heterogeneous. Such structures are identifiable because they are enwrapped with folded zones, and it raises a question of their similarity to the above-described inthrust massifs of the Okhotsk-Chukotka region. Such are (Fig. 38) the Indochina massif in Indochina which is enwrapped, in the northwest, with folded structures of Mesozoic age; and the entire South-China Platform together with a province of the caledonides identified in its southeast part by a number of researchers, which are to the south of the Ordoss trough. They are also enwrapped with folded structures on the northwest, being made up of geosynclinal Paleozoic deposits, but of Mesozoic age in origin. It is possible that a similar treatment is applicable to the structures of the North-China Platform and of the Zeya-Bureya massif together with the Hingan highland. As seen from Fig. 38, the above structures with boundaries of a parabolic shape form a nearly continuous chain, from the Aluchin structure (together, most likely, with that of the Great Anyui) to the Indochina massif.

On the probable nature of arch-shaped structures. Is it possible for such large structures to exist and move in a united manner under the effect of actively living seismofocal zones? The answer should be `yes' since the sizes of the arch-shaped structures identified in the above manner on the one hand, and seismofocal zones, on the other, are commensurable. Thus, a cross-sectional size of the largest structure among those mentioned—the South-China Platform—from the southeast to the southwest (together with the shelf) is about 2000 km; consequently, if it was restricted from the southeast by a paleoseismofocal zone at a depth of 700 km, then the latter's lowest base was beneath the central part of South-China (a structure at a distance of about 1000 km from the corresponding trench). The area of a force effect of the seismofocal zone might well extend to about the same distance inland. Some of the other structures identified above are nearly 2000 km in size or smaller (the Indochina massif is of about 1300 km from the southeast to the northwest, the Okhotsk massif is about 1000 km, the Omolon massif together with a part of the Taigonos structure is about 1200 km).

Obviously, the large arch-shaped structures described are rigid deep-seated massifs cut within hanging wings of seismofocal zones by pressure exercised onto these wings from beneath. An obligatory condition for their emergence is that their make-up should include rigid blocks of the ancient continental lithosphere, consolidated down to a depth of hundreds of kilometres. Thus, we can speak of depth structures subject to united tectonic displacements commensurable in size with the depth of seismofocal zones. Near the Earth's surface, their displacements take place as an inthrust of the rigid blocks of which they are comprised from the seismofocal zone side towards inland, if there is easily deformable lithosphere of geosynclinal origin; on the flanks, slip displacements develop from such inthrust structures, and at their front, overthrusting produces a bulging-up of their frontal edge. However, the main compensation for their horizontal displacements is due to crumpling of geosynclinal formations

Fig.38. Arch-shaped structures (inthrusts?) on Pacific margins of Asia. 1—Aluchin and 1a—possible Great Anyui, 2—Omolon, 3—Okhotsk, 4—Zeya-Bureyan , 5 -North-Chinese, 6—South-Chinese, 7—Indo-Chinese. 1—Arch-like fault zones, 2—frontal faults of Mesozoic Asian paleocontinent (upper border of continental slopes), 3—Talovka-Main line of Mesozoic eduction out-thrusting, 4—recent trenches, 5—strikes of folding structures.

in front of them. The existence of such structures and their movement are in accordance with the proposed model of eduction.

The horizontal inthrust of the arch-shaped massifs is smaller in size but undoubtedly may amount to hundreds of kilometres, which is confirmed by the values of crumpling in the geosynclinal areas. In the Aluchin structure, as was seen above, displacements of 100 km or more are identified with some confidence.

Reconstructions of the development of initial geosynclines in the North-East of Russia have made us repeatedly suggest horizontal displacements of their elements over some hundreds of kilometres. As shown, the possibility of such displacements in the hanging wing of seismofocal zones depends, under conditions of eduction out-thrust, on two facors. In one case, they are engendered by extension forces and accompanied by the opening up of a marginal sea or other extension structures; in the other case, they are caused by directed inland horizontal pressure due to eduction (Chapter III, Fig. 33) and accompanied by fold deformations. A condition necessary for such displacements is a heterogeneous structure of the hanging wings of the seismofocal eduction zones and the presence within them, in addition to rigid blocks of the ancient continental (platform) lithosphere, also of much more deformable geosynclinal matter. It should be supposed that it is the situation of the latter type that corresponds to deformations connected with origin of inthrust arch-shaped structures and folding of the Oloi-Anyui geosynclinal area in the middle of the Mesozoic.

DEVELOPMENT OF THE KORYAK GEOSYNCLINE

The Koryak geosyncline is distinctly bounded on the northeast by the Okhotsk-Chukotka land volcanic belt and in the southeast, its accompanying continental slope. The main (central) part of the geosyncline proper is represented by the area which is to the southeast of the linear Talovka-Main uplift, and comprises the folding structures of the Upper Jurassic and Cretaceous deposits. In the author's opinion, the Khatyrka Rise, which is to the south near the Bering Sea, and contains outcrops of Paleozoic rocks, should be considered to be a structural unit making up a part of the south margin of the Koryak geosyncline.

A border which separated the Koryak geosyncline from the Paleo-Asiatic continent was accompanied, as has been mentioned, by an ancient seismofocal zone, whose existence is recognized by many. If one assumes that it was similar to the present ones in size, then it was over it that the Okhotsk-Chukotka volcanic belt was disposed, and there was also a wider ore-magmatic belt which includes a `perivolcanic' zone. These were formed on a heterogeneous foundation, comprising massifs of the old continental lithosphere and different-age folded structures, including those inheriting the Oloi-Anyui geosyncline.

There is some uncertainty about the compression of the width and location of the seismofocal zone outcrop. Some researchers consider it to be a belt of up to 150-250 km width near the northwestern border of the Koryak geosyncline characterized by development of many large faults mainly parallel to this border; the zone thus identified spreads over a part of the volcanic belt on the northwest while its larger part corresponds to the Mesozoic continental slope of the Koryak geosyncline. To define this question, it seems necessary to isolate the surface of the main tectonic displacements (described in Chapter III) from the composition of the seismofocal zone which is treated as the location of possible occurrence of seismic foci in the past. In the present-day active provinces, the outcrop of such surface falls on the trench bottom and separates the crust of the continental or island-arc slope from the oceanic crust of the outer slope of the trench. Similarly, the site of the outcrop of this surface in the ancient seismofocal zones should be the border defined by the oceanic foundation on the one side and the continental-type foundation on the other. Such a border can be located quite definitely; it closely coincides with the Talovka-Main Rise—a peculiar linear structure of a horst nature stretching parallel to the volcanic belt and some 100-120 km to the southeast of it (Fig. 39). Many researchers have already pointed out that emerging in this structure and stretching farther to the southeast there is a typical ophiolite complex making up the basement of the Koryak geosyncline. However, this fact was not taken as a basis for immediate comparison of the Talovka-Main structure and the axial zone of the ancient trench. Meanwhile, the Talovka-Main Rise could, according to a number of its features, be considered to be a compensating structure which emerged in the place of the past isostatic negative anomaly which usually accompanies the trenches.

Talovka-Main Rise

The Talovka-Main Rise is a linear horst structure conspicuous in the recent relief and stretching over about 600 km from the Penjin bay in the southwest to the River Belaya in the northeast. Within different areas, it is occasionally expressed as a narrow tectonic range of 15-20 km cross-section and, in other instances, as a wider less distinct rise, but it is continuous everywhere. Most researchers consider the submeridional Pekulnei range to be a continuation of the Talovka-Main Rise to the north, that is somewhat displaced to the east. Including this range, the total length grows by another 100-150 km. On the two sides of the Rise, there are linear depressions parallel to it with a width of 50-70 km. A more uniform Penjina-Anadyr depression lies in the northwest and the Parapol depression, which accompanies the rise only in its southwestern half, is in the southeast. It may be supposed that there is a continuation of the Talovka-Main Rise far to the southeast on the bottom of the Sea of Okhotsk: this is concluded from the data of the deep seismic profile which established a buried horst-like linear structure in the sea basement south of Magadan (Zverev, 1964) over the strike of the Talovka-Main Rise, corresponding to its width. This suggestion seems to be reasonable, since the Talovka-Main Rise is one of the basic structural elements in the ancient marginal area which can stretch as far as other main elements, such as, for example, the volcanic belt, and others. If so, the length of the Rise is more than 1600 km.

Unlike the geomorphological structure, the tectonic structure of the Talovka-Main feature is very complicated. The structure concentrates in itself a vast amount of various dislocations with steeply and gently dipping contacts. Steeply dipping longitudinal faults bound the present-day horst structure and are widely distributed inside it: these are repeatedly described on the basis of geological mapping, and can be distinctly deciphered on airphotos. These relatively young faults cut the older more gently dipping ones: hence, the latter can be found only as fragments (as in the Ust-Belsky mountains, for instance). The decoding of the dislocations which are combined in such a way, and of the tectonic blocks formed by them, presents considerable difficulty.

The outcrops of the basic-ultrabasic rocks of the Koryak geosyncline basement which occur widely to the southeast of the Talovka-Main Rise spread into a number of localities within the latter, and occasionally occupy almost its entire width (for example in the Ust-Belsky mountains, or Pontonei mountains near the Penjina Bay coast). On the other hand, in some localities of the Talovka-Main Rise, blocks of Paleozoic deposits are common, among which are known faunally characterized Late Silurian, Early, Middle and Late Devonian, Early—Middle Carboniferous and Early and Late Permian deposits. These deposits of different ages make up mosaically alternating blocks, wedges and `lumps of indefinite origin' as Belyi (1963) puts it; in the localities of the Rise with outcrops of the ophiolite complex they are displaced to the northwestern side. The Paleozoic deposits differ greatly from the overlying Jurassic—

A
Pk
B
TM
Anadyr
Velikaya
Anadyr Bay
N
Pkh

Fig.39. Geological structure of Koryak geosyncline area and its continental Paleoasian surrounding. 1-4—Basement and sediments of Mesozoic paleocontinental Asian margin (of inshore and foreshore shelf and continental slope, including late sediments of Talovka-Main paleotrench): 1—supposedly Archean crystalline formations of Pekulnei Ridge; 2—Paleozoic deposits (Devonian, Carboniferous, Lower Permian); 3—basic and intermediate paleotypic effusions of possibly pre-Mesozoic age; 4—large fields of intrusive rocks (mainly granitoids, partly gabbroids), possibly including those of pre-Mesozoic foundation. 5-14—sediments related with paleocontinental slope: 5—Late Triassic, 6—Jurassic, 7—Neocomian, 8—Aptian-Albian, 9—Cenomanian-Turonian (partly continental), 10—Senonian, 11—Maastrichtian-Danian (with coal-bearing facies); 12—localities of immediate overlying of Barremian-Aptian or younger deposits on geosynclinal basement near axial part of Talovka-Main paelotrench; 13—prism of burial of Talovka-Main paleotrench, deposits accumulated after eduction movements stopped, of Barremian-Aptian, Albian, Cenoman-Turonian, and partly of younger age; 14—area of possible foot of Mesozoic continental slope on the side of Eskimo massif (Alkatvaam and Peculnei zones) with Jurassic, Neocomian, Aptiam, Albian, and Upper Cretaceous deposits. 15-18—Basement in inner part of Koryak geosyncline: 15—melange and micromelange (bold lines—localities of considerable distribution of serpentinite bodies and serpentinite melange), 16—especially large ultrabasic bodies; 17—their associated gabbroids, 18—detached fragments of surrounding continental foundation containing Silurian, Devonian, Carboniferous and Permian faunally characterizied deposits. 19-26—Sediments in inner part of Koryak geosyncline: 19—areas of known age of basal geosynclinal sediments in Late Jurassic-Neocomian interval: encircled are age designations of most ancient faunal finds contained in them: 1—Hauterivian, 2—Hauterivian-Late Valanginian, 3—Middle Valanginian, 4—Early Valanginian, 5—Berriassian, 6—Volgian (Tithonian); 20—Aptian-Albian basal sediments at south border of Velikorechensky trough; 21—Cenomanian—Turonian sediments in Velikorechensky trough, 22—mainly chert sediments of Coniacian-Santonian, 23—sediments of Santonian, Campanian, and Maastrichtian in Enchyvaam-Ukelayat basin, 24—Campanian intergeosyncline sediments connected with uplift at northern fringe of Velikorechensky trough, 25—Eocene sediments within the youngest geosynclinal basin (21-23 and 25— strikes of beds in fold structures); 26—part of Bering Sea in Oligocene and Neogene. 27-32—Fragments of outer Mesozoic island arc: 27—blocks of Hatyrka Rise with outcrops of Paleozoic deposits of Devonian, Carboniferous and Permian (including geosynclinal facies), outcrops of mafic Paleozoic geosynclinal basement and sedimentary covers of mainly Cretaceous age with inner discontinuities; 28—blocks of similar structure with no finds of Paleozoic faunal remains; 29—inner (facing the Koryak geosynclinal) slope of Hatyrka part of island arc (mainly deposits of Aptian, Albian, Senomanian, Turonian, occasionally Hauterivian); 30—intermontane and piedmont deposits of Maastrichtian, Danian and Paleogene; 31—blocks with Upper Paleogene and Miocene marine deposits of external slope of Hatyrka Rise; 32—Cenozoic and Upper Cretaceous sediments, mainly gently deformed, in Pachacha Block. 33-38—Plutono-volcanic structures of ore-magmatic activization areas, 33-36—within Mesozoic Okhotsk-Chukotka volcanic belt: 33—andesites and andesite-basalts of mainly Albian age, 34—liparites and dacites of mainly Senomanian, 35—effusions of variegated (contrast) composition, often rich in potassium, of Senonian age, 36—granitoid intrusions of mainly central type (partly subalkaline); 37-38—volcanic covers and extrusions of post-Senonian age displaced in fold area of the Koryak geosyncline and becoming younger, from Danian-Paleogene to Early Quaternary, southeast: 37—basalts and andesite-basalts, 38—liparite-dacites, dacites, liparites. 39-42—Tectonic contacts: 39—boundary of eduction out-thrusting during Jurassic-Neocomian established from observed changes in geological complexes and supposed; 40—upper (of frontal continental faults) tectonic restriction of Jurassic-Neocomian paleocontinental slope; 41—large overthrusts; 42—faults of different nature. TM—Talovka-Main Rise, Pk—Pekulnei Ridge, A—Anadyr Fault, B—Berezovsky Fault, Pkh—Pakhacha Block.

Cretaceous ones in the peculiarly irregular conditions of their occurrence and also due to ubiquitous stratigraphic discontinuity. Some of the Paleozoic deposits, which are of various facies, can be correlated with similar-age formations known to the northwest in the structures of the Paleo-Asiatic continent, in particular in the Yeropol Rise and adjacent parts of the Omolon massif.

It should be noted that the Paleozoic blocks and lumps, on the one hand, and the ophiolite complex, on the other, make up the foundation of the Talovka-Main Rise zone overlain by the Mesozoic sediments. We should relate the crystalline granulite-basite complex of rocks in the Pekulnei Range to the same foundation. Markov *et. al.* (1982) cited convincing arguments in favour of this complex possibly belonging to the Archean formations. The overlying Mesozoic deposits of the predominantly Middle—Upper Jurassic and Cretaceous age should be viewed as a sedimentary cover of the continental slope and a filling of the trench which came into being in the process of the burial of the latter. Whereas the ophiolite group of rocks undoubtedly belongs to the geosynclinal basement, the Paleozoic and Pre-Cambrian blocks probably are fragments of the paleocontinental slope foundation. This conclusion is confirmed by the composition and genetic features of the Mesozoic Jurassic—Neocomian deposits covering these blocks. The Talovka-Main structure on the whole is a 'counter-rise' formed instead of the trench, which thus includes the parts of both oceanic and continental foundation (which is natural because the axes of the negative isostatic anomalies are usually displaced towards the continental slope with respect to the trench axis).

The gently sloping large faults in the Talovka-Main Rise and in the Pekulnei Range are connected with the serpentine packets, altered gabbroides and melange. The formations of this kind are considered to be traces of large overthrusts and overthrust sheets. It is quite reasonable, however, to take them for structures of eduction out-thrust at the foot of the ancient continental slope on its border with the oceanic crust. In many localities, such border can be drawn quite distinctly; it divides, on the one side, all the outcrops of the ophiolite complex and, on the other side, the pre-Mesozoic rocks of the possible continental foundation together with their covering sediments of the paleocontinental slope. In the Talovka-Main Rise, the ophiolites are common to the southeast of this border, and the border itself is distinctly inclined together with numerous tectonic flakes and slabs dipping to the northwest, i.e. towards the paleocontinent (which were described by A. A. Alexandrov, 1978, and other geologists). Such a position of the border corresponds to the out-thrust model.

For the Pekulnei Range, researchers (Markov *et al.*, 1982) point out mainly the opposite inclination of the scales. The tectonic situation is very complicated and it is probably too early to draw a conclusion on the initial inclination of the sloping tectonic contacts; however, the opposite, i.e. eastern, out-thrusting of the depth matter within the Pekulnei Range seems quite possible owing to an obvious turn of all the structure of the Mesozoic margin area, and in particular of its volcanic belt, to the southeast which is observed to the north from the Pekulnei Range. The eduction out-thrust in this case must have occurred from the northeast, also from beneath the continental slope, but from the side of the Eskimos massif. The occurrence of the geological complexes in the Pekulnei Range also allows for this possibility.

In the middle section of the Talovka-Main Rise (the southern part of the territory in Fig. 39) the border of the eduction out-thrusting seems to be displaced beyond the boundaries of the present-day horst (which is very narrow here) deviating away from it to the southeast.

The northwestern paleocontinental slope

The age of sediments on the paleocontinental slope is of special significance in clarifying the history of the adjacent geosyncline. The continental slope is a stationary part of the active marginal area which bears a reasonably complete section of deposits corresponding to the time of existence of the geosyncline basin. In contrast, the sedimentary section of the eduction geosyncline where it adjoins this slope is incomplete (Chapter III; Fig. 34, also in the subduction case).

The area undoubtedly belonging to the continental slope of the Koryak geosyncline occupies a band of a constant width of an order of 100 km, and includes a number of longitudinal faults of the northeastern direction. Its northwestern boundary is the southeastern border of the land Okhotsk-Chukotka volcanic belt. This band is close to the continental slopes of the present-day trenches in width. The major part of its southwestern half is adjacent to the

Talovka-Main Rise, and is buried under the Neogene—Quaternary deposits of the Penjina-Anadyr depression, but in its northwestern part, the outcrops of the Mesozoic sea deposits are considerably spread out (Murgal Rise and adjacent areas). They are represented by sandstones, grits and sedimentary breccia of the Lower Jurassic, aleurites, sandstones and grits of the Middle Jurassic (in the northwest of the Pastbitschny Ridge), sand-and-aleurite sediments of the Berriasian, Valanginian and Hauterivian in the Murgal Rise, and of rough clastic deposits of the Barremian (which are usually identified conventionally), Albian, Cenomanian, Turonian and Senonian, the latter being mainly continental. Outcrops belonging to the basement of the paleocontinental slope on these areas are absent, with a possible exception only of some intrusive fields. The fold deformations are moderate but in the upper parts of the Albian the angular unconformity is distinct. The section of the paleocontinental slope thus testifies to the existence of the adjacent sea basin during the Jurassic and Neocomian and to the restructuring of the territory of the slope at the end of the Early Cretaceous.

The deposits of the Middle and Upper Jurassic, Berriasian, Valanginian and most Hauterivian which outcrop in the Talovka-Main Rise and cover there the blocks of the Paleozoic rocks in many cases should be considered to be a continuation of the sedimentary cover of the paleocontinental slope. The difference is in two- to three-fold smaller thickness of the sediments within the Talovka-Main zone and considerably more widely distributed fine-grained and cherty varieties of the same. These peculiarities fit the conclusions concerning the deposition of the Jurassic-Neocomian sediments in the Talovka-Main Rise near the foot of the continental slope in the deep-water trench. Some researchers (for instance, G. P. Terekhova, who studied these deposits thoroughly) point to the unambiguous signs of their deep-water origin.

The overlying deposits from the Barremian to the Upper Senonian in the Talovka-Main Rise are of coarse-grain nature and greater thickness, growing more shallow-water upwards along the section. Their accumulation obviously signifies the filling-up and burial of the paleotrench. In spite of their complex tectonic structure they and the underlying Neocomian deposits were very thoroughly studied by G. P. Terekhova, G. I. Agaltsov, V. F. Belyi and other researchers. The distinct unconformities are associated here not only with the Upper Albian, but also with the beginning of the Coniacian and Maastrichtian. Starting from the end of the Albian, in the sedimentary section of the Talovka-Main Rise, there emerge boulder conglomerates and lump formations of an olistostromes type with fragments of the Paleozoic sediments and granitoides. One can suggest that this time is associated with the early movements which initiated the compensating uplift in the trench zone. However, the main upward movements took place probably as late as the end of the Senonian with the trench already filled up with sediments: this can be seen from a sharp bending of the Senonian deposits (including the Campanian ones) near the western border of the Talovka-Main Rise (see Fig. 39). Early clastic fragments of the ophiolite complex, including ultrabasites, emerge in the Early Senonian sediments of the Pekulnei Range.

Thus, proceeding from the data obtained in the zone of the northwestern continental slope of the Koryak geosyncline, this zone developed actively, with the deep-water trench present in its northwestern part, during the late Jurassic and Neocomian. Let us compare this conclusion with the data obtained in other ways.

The spatial interrelation of the Koryak geosyncline with more ancient systems

The northwestern margin of the Koryak geosyncline was represented by the edge of the Paleo-Asiatic mainland disposed to the northwest of the present-day Talovka-Main Rise. The mainland foundation was made of the totality of large massifs — fragments of the ancient platform lithosphere — and folding structures which emerged to take the place of the Paleozoic—Early Mesozoic Oloi-Anyui geosynclinal system.

As it was shown, it is possible to establish for the Oloi-Anyui system the following: 1) the system's main structures are of a distinct northwestern strike and join, sharply unconformably and according to the end type, with the border of the Koryak geosyncline of the northeastern strike; 2) the system's development was ended in the middle of the Mesozoic with a folding which consistently spread over its different parts: in the Triassic, including the Carnian and Norian age, sedimentation still occurred there in large geosynclinal basins, but during the Early and Middle Jurassic the basin areas reduced considerably; during the Late Jurassic and the beginning of the Cretaceous, the folding affected the whole geosynclinal area excluding only a part of the South-Anyui zone; during the Hauterivian—Barremian, that last basin was eliminated by the folding process too.

Development and filling by sediments of the Koryak geosyncline basin took place from the Jurassic to the Paleogene. Thus, neither the structure strikes, nor the time of the sedimentation and folding share any common features in development with the two contiguous geosynclinal provinces of the Koryak and Oloi-Anyui. On the contrary, the sequence of events makes one think that the emergence of the former province caused the closure of the latter: following the beginning of sedimentation on the continental slope of the Koryak geosyncline there was fold transformation of the Oloi-Anyui system. The interrelation of the strikes indicates the superimposed nature of the Koryak structures. It would be logical to suppose that the superposition of the Koryak geosynclinal system upon the Oloi-Anyui was expressed originally as a linear break up of the Earth's crust transverse to the strike of the Oloi-Anyui geosyncline. The emergence of the eduction seismofocal zone inclined to the northwest that followed caused folding in its hanging wing, within the structures comprising the Oloi-Anyui geosynclinal system (according to the mechanism described in Chapter III), which initiated the further growth of southeastern margin of Asia.

Time of the opening

An opinion which in some cases asserts that the Koryak geosyncline emerged and existed during the Paleozoic or even earlier is based on: 1) interpolation of the Paleozoic rock outcrops, which are known to exist in the western and eastern frame of the Koryak geosyncline, into the area of its central part; 2) the presence of small isolated but numerous tectonic blocks of the partly faunally characterized Paleozoic rocks in the basin of the rivers Vayega, Mukarylyan, First and Second Berezovaja at a distance of up to 70 km to the southeast of Talovka-Main Rise: 3) age dating of metamorphic rocks of green schist facies which occur over a number of localities in the central part of the geosyncline as conventionally Paleozoic; 4) suggestion that each and every point of the Earth's present-day surface has always had its corresponding ancient Earth's surface.

All the above arguments can be disputed, though the last one, as applied to the Koryak geosyncline, might be considered reasonable in the case that it emerged owing to the basification of the continental crust or it inherited the ancient protogeosyncline. Such suggestions were made, for instance, by Alexandrov (1978). However, the rupture of the Earth's crust, which obviously discontinued the northwestern structures of the Oloi-Anyui system and started developing the Koryak geosyncline, could hardly be thought of as other than of a fault nature, and one that produces a thrusting apart. An opinion on the thrust-apart nature of the geosyncline's opening is now supported by a number of researchers adhering to various tectonic positions; the more so as it relates to the opening of the present-day marginal seas (e.g. Kropotkin, 1980; and others) which are justly considered by many tectonists to be analogues of ancient geosynclines. With thrust-apart origin of the geosynclines, the mechanism of formation of their foundation can be of a spreading nature and, according to the author of this work, of an eduction nature as well. In either case, the basement is created anew, i.e. forming a new Earth's surface. Therefore, for the Koryak geosyncline, as well as for any other specifically treated geosyncline, one should identify its surround (or frame) and the inner part created as a result of it being thrust apart. Thus, it is not permissible to interpolate into the thrust-apart portion of the geosyncline data about the age of sediments at its outer frame, if they are more ancient than the time of the geosyncline opening (or it is impossible to raise a question of what had taken place in one or another point of the thrust-apart section of the geosyncline before that, since no corresponding surface had as yet existed on the Earth).

An idea of the existence of the Koryak geosyncline in the Paleozoic does not correspond, first of all, to general features of the geological structure which characterize this geosyncline. Within the geosyncline, we can study continuous sedimentation (with local complications, interruptions, and displacements of sedimentation areas to be discussed below) starting from the Upper Jurassic through the Neocomian, Aptian—Albian, all the divisions of the Upper Cretaceous and up to and including the Eocene. No doubt, these sediments correspond to the time of existence of the Koryak geosyncline basins. One could suppose that sedimentation might have begun immediately preceding the Upper, Middle and Lower Jurassic epochs, perhaps in the Upper Triassic, if one suggests that they are represented by barren facies and are of limited distribution. However, it is quite obvious that all the Paleozoic deposits (mainly Devonian, Carboniferous and Permian), which are found in the geosynclinal surround or revealed within inner parts of the Koryak geosyncline, are separated from the continuous Mesozoic stratigraphic section by discontinuities accompanied by differences in metamorphism, facies, etc. If one supposes that in the place of the Mesozoic Koryak geosyncline there existed an earlier Paleozoic geosyncline, then the latter had obviously undergone a folding and other transformations after which the Mesozoic geosyncline (as a geosynclinal basin) emerged anew. However, there are not really any grounds for believing in this suggestion.

One should relate to the foundation of the Koryak geosyncline not only the ancient blocks which are Archean in their base, but also the areas of distribution of the Paleozoic sediments, such as the slopes of the Omolon massif. The geosynclinal zones of the Oloi-Anyui system also became a part of the surround (at the northwestern continental side) of the Jurassic-Cretaceous geosynclinal Koryak basin. Earlier they developed during the beginning of the Mesozoic and were positioned among the ancient massifs of the Archean and fold formations of Early—Middle Paleozoic age, but were folded at the end of the Triassic or the beginning of the Early Cretaceous, together with more ancient tectonic elements. So the marginal foundation included not only Paleozoic but also Lower Mesozoic folded formations was superposed by the initial Koryak structure with a new northeastern strike, and later affected by thrusting apart. In particular, the outcrops of the Paleozoic rocks in the Talovka-Main Rise are usually considered to be fragments of the continental slope foundations which were affected by the block displacements.

A large structure, which includes Paleozoic rocks, is the Khatyrka Rise and its adjacent areas. Here, there are Paleozoic deposits of different ages, partly with geosynclinal facies, accompanied in some cases by ultrabasite outcrops. The Paleozoic rocks in the Khatyrka Rise are overlain partially by the Neocomian and partially by Upper Cretaceous deposits up to and including Maastrichtian. This allows treatment of the Khatyrka structure as one with a Paleozoic basement which underwent repeated elevations during the sedimentary filling of the Koryak geosyncline; it thus made up the southeastern surround of the main geosynclinal Basin which was pronounced in relief. All this is confirmed by outcrops of metamorphic rocks in the basin of the River Iomrautvaam (with contribution by amphibolites and mica schists, occasionally with garnet).

It is well-known that the structure of the Khatyrka Rise, and of its possible western continuation (Fig. 39), is very complex. Within the rise, one can observe a series of tectonic blocks, wedges, overthrusts and nappes (e.g. Alexandrov, 1978;, Ruzhentsev *et al.,* 1979; Markov *et al.,* 1982; Bragin *et al.*, 1987; Sokolov, 1990; Grigor'ev *et al.,* 1995). Along with the Paleozoic deposits represented both by facies of a geosyncline type (cherty schists, phtanites, clayey-aleurite deposits of deep origin) and shallow-water organogenous limestones; the serpentine melange complexes are also widespread within the rise. The presence of the geosynclinal Paleozoic deposits allows the supposition that some melange formations are also Paleozoic. On the other hand, contacting the Khatyrka Rise, there are to the north of it the fields of the Mesozoic melange formations of the central Koryak geosyncline area which makes it possible that its blocks are also present among the Paleozoic. Thus, decoding of the initial structure of the Khatyrka Rise is a rather challenging task. However, the existence in the eastern part of the Koryak geosyncline of the Paleozoic blocks of the surround's foundation detached from the northwest remains possible, one reason being that there must have been a structure (of island arc type) which separated the Koryak geosyncline from the rest of the open Paleo-Pacific Ocean. Probable remains of such a structure are now observed as some relatively rigid blocks near the southeast border of the central area of the Koryak geosyncline (the Khatyrka Rise is one among them), as well as to the south of it (the Mid-Kamchatka massif of metamorphic rocks). The continuation of this structure of island arc type is probably partly hidden under the Sea of Okhotsk.

Thus, distribution of the Paleozoic deposits in the Khatyrka blocks, as well as in the other similar ones, cannot be considered a proof for the existence of a Paleozoic geosyncline at the location of the Mesozoic Koryak geosyncline. However, the above mentioned block outcrops of Paleozoic rocks in the basins of the rivers Vayega, Mukarylyan and Berezovayas cannot be related to the outer surround of the Koryak geosyncline but belong, undeniably, to its inner part.

The area of the basins of the above rivers is very peculiar. The Paleozoic deposits among which there are faunally established the Upper Silurian, Lower, Middle and Upper-Devonian, Lower and Middle Carboniferous and Upper Permian (i.e., the same as are known within the Talovka-Main Rise) make up irregularly dispersed isolated tectonic blocks of a size from dozens or hundreds of metres to a few kilometres. These blocks are called `exotic cliffs', tectonic detached masses, or klippens (Kaigorodtsev, 1963; Chekhov, 1982). The `cliffs' are occasionally found in the fields of the sand-aleurite deposits of the Hauterivian or Valanginian-Hauterivian age as well as among difficult-to-distinguish tectonized masses of essentially mafic composition which are arbitrarily dated in different localities as Pre-Cambrian, Carboniferous, Permian-Carboniferous and Valanginian. These blocks have undoubtedly been displaced far from their site of formation and thus cannot be used as proof of the Paleozoic age of their entire surroundings. Eductionally, they should be considered as detached masses from the foot of the paleocontinental slope near the place of out-thrusting depth material onto the Earth's surface (i.e., in the zone of the Talovka-Main Rise) and transported into the geosyncline area proper together with the eduction basement. It is not ruled out that some of these detached masses were created by land from the paleocontinental slope; this supposition is prompted by the occasionally observed partial or even complete enveloping of the Paleozoic blocks with the terrigene deposits of the Upper Triassic age: these are also separated from the surrounding Neocomian deposits by contacts which are nonconformal to the sediment's bedding and so mapped as tectonic (in particular, G. G. Kaigorodtsev (1963) described the inclusions of many-metre lumps of Paleozoic limestones in the Upper Triassic deposits; such interrelations as well as the instances when the Paleozoic deposits are overlapped with the Upper Triassic formations were also recorded by A. D. Tsvetkov, Er. B. Nevretdinov, and others). Accordingly, the conventional dating of barren metamorphosed schists as Paleozoic in the Vayega area basin and some other places is also groundless.

Thus, there are, in the author's opinion, no convincing reasons to consider the Koryak geosyncline as a Paleozoic structure. If one treats its historical development as the opening of a thrust-apart basin, one arrives at the conclusion that this basin emerged and developed as late as the middle Mesozoic. This conclusion is confirmed by: 1) a sharp superimposition of the Koryak geosyncline upon the structures of the early Mesozoic Oloi-Anyui system; 2) the post-Triassic (possibly with the late Norian) age of the deposits in the cover of its Paleo-Asiatic continental slope; 3) the absence of reliably dated deposits older than Late Jurassic in the inner part of the geosyncline (excluding the erratic masses) with their overlaying of the ophiolite basement (Nekrasov, 1976; Alexandrov, 1978; Ruzhentsev *et al.*, 1979; Markov *et al.*, 1982 and others). The most probable time when the initial breakup occurred, that cut the previous structures and later developed into the geosyncline, is Upper Triassic.

The form of opening

Eductionally, the basement of the Koryak geosyncline formed by its out-thrusting in the Jurassic-Cretaceous Talovka-Main trench. On the side of the Pacific, the new basin which originated at the end of the Triassic or the beginning of the Jurassic and superimposed very unconformably upon the earlier Oloi-Anyui geosynclinal system, was bounded by the island Hatyr-Kamchatka chain — an island arc whose foundation had detached from the northwestern surround of the Koryak geosyncline when it was initiated. The composition of this foundation, as well as of the northwestern margin proper, should thus have been heterogeneous. Whereas the southwestern and central parts of the arc were positioned opposite the Okhotsk and Omolon massifs of the Archean-Early Proterozoic rocks (in the locality of that part of the Koryak geosyncline which is now under the Sea of Okhotsk), the composition of the arc foundation was contributed by detached blocks of the ancient continental lithosphere (the Eastern Kamchatka), and the northeastern part of the arc was positioned opposite the Oloi-Anyui geosyncline structures. These consisted of the former geosynclinal troughs of the Paleozoic and Early Mesozoic age alternated with more consolidated lithosphere of the Yarkan-Nutesyn Rise type. As all these structural zones were of a distinct northwestern strike, they are at present discontinued abruptly by the Koryak geosyncline. Structurally, their displaced continuation must also exist on the other side of the geosyncline . The place to search for this continuation, both spatially and compositionally is the Khatyrka Rise. If the supposition is true, then this rise could separate from the northwestern surround of the geosyncline with an incomplete closure of the previously existing geosynclinal trough of the Oloi-Anyui province, which caused its especially complex structure and dislocations.

If, in front of the Koryak geosyncline, there existed its bounding island arc, there should also have existed an eduction seismofocal zone beneath it, with outcropping onto the surface as the deep-sea trench on its oceanic side. In this case, the Koryak geosyncline was a marginal sea separated from the ocean by an island arc and bounded by two seismofocal zones—one on the internal border where the Talovka-Main Rise is now located, and the other on the outer border of the arc. The structure of the Koryak Mesozoic geosynclinal basin proves, thus, to be similar to that of the present-day Philippine Sea. Obviously, if such a double system of one-side-inclined eduction seismofocal zones existed, the outflow of depth material from the outer zone, facing the open Pacific, caused separation of the island arc from the continent and the opening of the marginal sea behind it. The other eduction zone (in this case, the Talovka-Main one) provided the advent of the matter to build the basement of the opening marginal sea; it was, thus, of secondary character. Within the area between the Talovka-Main and Khatyrka Rises, in the Koryak geosyncline, the above-described model stipulates ubiquitous sediment-underlying ocean basement of eduction origin, with the age of overlying sediments consistently growing from the Talovka-Main zone (except for sediments of the paleocontinental slope) towards the Khatyrka Rise.

Eductional basement

Within the course of a few dozen kilometres of its journey before emergence onto the Earth's surface, the matter of the out-thrusting wing near the eduction zone is in continuous interaction with the material of the stagnant continental wing, destroying the latter and being itself destroyed, in the process of cooling and solidification of the material of the out-thrusting wing. The transverse sizes of the destructive zone along the eduction surface, with the amplitude of relative movements of hard masses of tens to hundreds of kilometres, and judging from the known thicknesses of the tectonic products in the faults of more moderate amplitudes, should be of the order of several kilometres. Some products of tectonic destruction are inevitably ejected onto the Earth's surface during the eduction out-thrust process; thus, the geosynclinal, i.e. oceanic, crust of eduction origin should bear in its upper near-surface part a cover of the tectonic formation of mainly mafic, ultrabasic and basic composition, which corresponds to the mantle and lower parts of the continental Earth's crust material, and partly also, to the granite continental layer's rocks. According to the author's observation, within the Koryak geosyncline, there is indeed distributed such material, and one can identify three main varieties of such eduction basement: melange, micromelange and low-temperature metamorphic schists.

Melange occurs over different parts of the Koryak geosyncline and has been described by many authors (e.g. Alexandrov, 1978; Chekhov *et al.*, 1980; Alexeev, 1982; Markov *et al.*, 1982; Sokolov, 1990; Silantiev *et al.*, 1994; Grigor'ev *et al.*, 1995). It is a tectonic mixture of frequently variegated composition which is often essentially serpentine. It contains randomly dispersed angular and tectonically rounded-off inclusions which vary in size from small fragments to large lumps and occasional tectonically detached pieces of hundreds of metres and larger. The fragments and lumps are represented by serpentinized to a variable degree ultrabasites, gabbroides, metamorphic rocks of various facies, from eclogites to green schist (including glaucophane-lawsonite schists), and also examples of Paleozoic formations of different age and origin. The formation of melange is complex and in different places can be dissimilar. The origin of melange is most frequently attributed to overthrust and nappe dislocations which developed in the course of folding and accompanying processes. However, melange in some places is clearly overlain by Upper Jurassic and Neocomian deposits

(Alexandrov, 1978; Ruzhentsev *et al.*, 1979; Markov *et al.*, 1982). Consequently, it made up the floor of the opening geosyncline as early as the beginning of sedimentation and long before folding processes developed in it. This is in fair agreement with the suggestion concerning its eduction origin. A wide distribution of melange in itself, represented by mainly ultrabasic rocks, testifies to the elevation of large masses of the mantle levels to the Earth's surface which is the essence of the eduction process. Distribution of the melange in the Koryak and other geosynclines is widely covered in the literature, but the question of its existence there and in the basement of recent oceans was raised by the author of this book not very long ago (Chudinov, 1979, 1981).

Micromelange. Within the Koryak geosyncline there are various localities that are distributed over large area, which are sites of fine-grained rocks of mafic composition; the nature and origin of these rocks is difficult to determine. These rocks are defined as altered basalts, spilites, diabases, especially frequently as altered tuffs, tuffites, tuffisites, etc., most of them having completely changed their initial textures. They are usually easily hand-crushed fine-fissured dark-coloured, green and gray, rocks bearing along their fissures many glide planes and curved surfaces. It is significant that they have a network of fine irregular veins of zeolites (often of analcime), calcite, occasionally prehnite, chalcedony, talc, etc. Microscopically, the rocks reveal a mainly fine-grained, almost opaque, mass wherein some localities can be seen with bits of parallel structures comprising an aggregate of fine-scaled chlorite, partly serpentinite; one can sometimes discern also grains and aggregations of actinolite and other secondary hornblendes, epidote, opaque ore minerals, zoisite, etc. Such formations, as a rule, are attributed as much altered volcanogenic rocks, on the basic of their melanocratic composition, areal extent (which is usually not supposed to be common to tectonic products) and their occurrence in many cases near the base of sedimentary sections. Determination of age of overlying or closely-occurring sediments also usually governs the dating of such rocks; in different localities of the Koryak geosyncline this dating depends on the surrounding conditions and is taken, for example, to be pre-Cambrian or Early Paleozoic near the detached bulks of Paleozoic rocks in the Vayega region and over some localities of the central part of the geosyncline (the upper reaches of the River Belaya, the basin of the River Lesnaya and some others, where the age was estimated from the degree of metamorphism), or Late-Jurassic and Neocomian (most often), or Campanian and Maastrichtian as it mostly is admitted in the southern part of the geosyncline.

Study of the rocks of the above type over different localities of the Koryak geosyncline, as well as within the Oloi-Anyui geosynclinal area (where their dating differs but the character of rocks is the same) led the author to the conclusion that they are worn-through ground material of tectonic origin, of various, but mainly basic and transient-to-ultrabasic composition, i.e. similar to the material of the basic mass of many varieties of melange. These are mafic cataclasites and mylonites of very wide distribution. The ultrabasic material seldom contributes to their composition since, when worn-through, it is mainly of serpentine and is diagnosed as serpentinites which are easily recognized; but numerous serpentine bodies of various size, indeed, always accompany the micromelange fields of any considerable size.

Being attributed to volcanogenic formations, micromelange is included into stratigraphic units together with its overlaying or closely located sedimentary rocks. In accordance with the composition of sediments, such masses are described as volcanogenic-sedimentary, flinty-volcanogenic (due to the cherty sediments which are often observed to be widely distributed in the overlaying deposits), tuff-cherty-terrigene, etc. An obvious contradiction in identifying such stratigraphic units is the sharp contrast between the supposedly deep change in the volcanogenic rocks but insignificant metamorphism of the overlying sediments that are easily diagnosed.

Rocks such as those described above are well known to geologists who work in the Koryak geosyncline area. There is even a common provisional name, `shukhobit' or `shukha', derived from a word of a regional dialect (a residual product in making sunflower oil, as told to me by A. D. Chekhov); this word was used conversationally and in field descriptions to designate unidentifiable disintegrated mass, which is mainly basic in its micromelange composition, but partially represented by its accompanying heavily cataclized aleuro-sandstones, cherty schists, and so on. The appearance of the latter rocks in the fields shared by it with micromelange is quite explainable, since micromelange is a weak, easily deformable mass, which is the reason why its overlying rocks turn out also to be deformed.

Low-temperature metamorphic schists of green-schist facies are often found to be intimately connected territorially with formations of micromelange type, with gradual transitions existing between the former and the latter. In different localities of the Koryak geosyncline territory there occur amphibolite-chlorite, actinolite, epidote-amphibolite, albite-epidote-actinolite and other types of schists including some containing glaucophane and lawsonite (in the ridges of Pekulnei, Vaezhsky, Chirynaisky, and other regions). The latter are usually attributed to the facies of `blue schists' which are considered to be typical of subduction zones. However, the conditions of origin reflected by glaucophane and lawsonite — high pressures with relatively low temperatures — equally correspond to the process of eduction out-thrusting; for the latter accounts for a mass delivery to the Earth's surface of products of such metamorphism (Chapter II).

The totality of the above-mentioned formations make the bulk of the known outcrops of the Koryak geosyncline basement. A feature of the structure of the Koryak geosyncline (which seems to be true of other geosynclines of similar origin) is a sharp boundary between the sedimentary cover and the basement they overlie, if one includes also micromelange with the latter (since micromelange is usually described as a volcanogenic rock). Interbedding of the supposedly volcanogenic rocks of this kind with the sedimentary ones is mentioned infrequently and each time may be doubted because of the complex tectonic structure of the corresponding areas, as tectonic alternation of the sedimentary rocks with, for example, large-fragmental melange is well-known (e.g. Alexandrov, 1978; Ruzhentsev *et al.*, 1979). Nevertheless, interbedding zones are described near the boundaries between the larger fields of the basement and the sediments; as a rule, there are also no descriptions of volcanogenic horizons in the upper parts of the sedimentary sections, except for some quite reliable tuffs. (An exception is the basin of the River Pikasvayam, on the left-bank area of the River Ukelayat, where the existing maps by V. E. Pinte show alternation of two thick series of altered volcanogenic basic rocks and two series of barren sediments overlying them, dated of the same Campanian age from comparisons with the neighbouring areas. It is possible that here is a repeated section of the same sediments with their eduction basement).

It is thus typical of the Koryak geosyncline to exhibit the same interrelation between the sedimentary cover and the basement that is found by deep-sea drilling in the recent marginal seas and everywhere in the ocean areas adjacent to the trenches: these are sharply divided, and there is no intermediary interbedding zone between them.

The only area within the Koryak geosyncline known to the author (with the exception of the Khatyrka blocks from it, according to what has been said above) where there are developed covers of reliable volcanic rocks is the basin of the River Nauchirynai (the left tributary to the River Velikaya) south of the Zubchaty Ridge. These rocks of basic composition (andesite-basalts) are unequivocally established from the porphyritic textures common to lavas); they overlie here the serpentine melange of the Yagel zone (according to A. A. Alexandrov) and are, in turn, overlain with terrigene Valanginian deposits. On the whole, however, distribution of reliably identified volcanic flows over the territory of the Koryak geosyncline is rather insignificant.

Age of the sedimentary cover base

During the eduction out-thrust of the geosyncline basement in the trench disposed on one side, its sedimentary cover becomes of asymmetric structure: the age of the base of the sediments covering its basement gets increasingly older farther from the trench, and the age range of the entire sedimentary cover grows (Fig. 34). As long as the eduction mechanism works, the trench itself accommodates only layers of the youngest sediments (which can be observed in the recent trenches too). The facies of sediments deposited in the trench and then transported towards the open part of the geosyncline, thus propagate due to eduction over the entire geosyncline area, with their age varying depending on the distance from the trench.

In most cases, uniform grey carbonate-free sandstone-aleurite deposits, which are either massive or layered and occasionally of a gradation type, prove to overlie the tectonized basement of the Koryak geosyncline, or those distributed in the nearby fields. Over considerable areas, these deposits contain no fauna, which makes it very difficult to place them chronologically; yet there are some localities where they are characterized faunally, normally with just a few species of pelecypods and ammonites.

It is commonly believed that in the northwest, in the zone of the Talovka-Main Rise, the lower constituents of the sedimentary section overlying the supposedly volcanogenic deposits (i.e. in many cases, micromelange) are made up by Valanginian deposits. The latter and the rocks of the micromelange group are often included into the structure of the Pekulneiveem suite dated as Valanginian, or as the Upper Jurassic—Valanginian on the basis of comparison with the eastern portions of the Koryak geosyncline. Dating them as of the Valanginian is based on observations made in the Ridge Pekulnei, where the rocks of the micromelange group outcrop onto the surface near the terrigene deposits with Valanginian fauna (with no contacts directly observable). The finds of the Berriasian-Valanginian related to the same Pekulneiveem suite are also found in the Talovka-Main Rise; however, these localities are disposed near the border between the former foot of the paleocontinental slope and the oceanic basement of the geosyncline. This is why the deposits outcropping onto them can belong to either the former or the latter, and hence, it is certain that the Neocomian deposits contribute to formation of that paleocontinental slope cover. To solve the question of the age of sediments overlying the geosyncline basement in the trench area, one has to make direct observations ruling out the possibility of attributing the stratigraphic significance to tectonic contacts, and such observations have indeed been made.

In the upper reaches of the River Vesnovannaya in the southern half of the Pekulnei Ridge, V. I. Shkursky and G. P. Terekhova have found and described a complex of flinty and green-schist rocks with serpentine bodies (related to the volcanogenic part of the Pekulneiveem suite) overlying (over several kilometres) by faunally characterized as Barremian—Aptian sediments with a constant pebble horizon at their base. The Barremian—Aptian deposits are overlain by the Albian—Turonian ones and then by younger Upper Cretaceous deposits of clastic and tuffaceous composition. Obviously it is this superposition of sedimentary cover upon the ophiolite-micromelange complex that indicates the time when the recently observable sedimentary mass began to accumulate on the oceanic basement proper of the Talovka-Main zone.

To the south of the Pekulnei Ridge, on the eastern slope of the Talovka-Main Rise in the basin of the River Utesiki and farther to the south where there are numerous outcrops of the typical micromelange and melange (Alexandrov, 1978; Chekhov, 1982; and observed also by the author), there is an intricately dislocated sandstone-aleurite series, also containing grit and conglomerates, which was described as Algan series. It also overlies the supposedly volcanogenic formation (treated by us as micromelange). The sediments are poorly characterized faunally, but in its southern outcrops, V. F. Belyi and V. G. Kal'yanov have collected Cenomanian—Turonian fauna (in the former case, it was initially regarded as Senonian but later re-identified). In the lower reaches of the River Chivetykveem, Kal'yanov also identified fauna of the Later Valanginian, but judging from the descriptions, it was contained in some exotic lumps, possibly similar to the detached blocks of the Mukarylyan — Vayega field: these are several skerry outcrops of fine-laminated jaspers, differing sharply from the surrounding sandstone and clayey slates on the watershed of the River Pravyi Konachan.

If one considers these fauna to be introduced accidentally into fragments of older origin, then, according to the rest of the fauna, the Algan series should be dated as Cenomanian—Turonian with possibly an older age (the Aptian—Albian?) for its lowest part. The southwestern continuation of the Algan series probably is a band of sandstones and sandstone-aleurites characterized by rare Aptian—Albian fauna discernible on the left-bank area of the River Main as far as its upper reaches. The totality of deposits, found at the above localities from the Pekulnei Ridge, make up a united band extending for 400 km, which can be taken for a single structural-formation zone — a prism of sediments which filled up the trench after the out-thrust eduction process stopped, its base being of the Barremian—Aptian age.

This conclusion is compatible with the data on the age of sediments which overlie the eduction basement of the Koryak geosyncline, if one traces them southeastward across the strike of the geosyncline. To the southeast from the area of development of the Algan series, in the basin of the Rivers Vayega — Mukarylyan — Berezovayas, there are terrigene deposits characterized by Hauterivian fauna; these deposits overlie the dynamo-metamorphosed schists, melange, and Paleozoic and Late Triassic rocks found as exotic lumps. Identified here also are the deposits of the Valanginian, whose age, however, is established from the presence of unidentifiable faunal remains detected together with the fragments of *Inoceramus* shells (the River Vayega basin). The *Inoceramus* in the Early Cretaceous (according to Terekhova and others) appeared in the Koryak geosyncline as late as the Hauterivian (earlier, there were only *Buchia*). So the base of the sedimentary section in the Vayega area should most probably be related to the Hauterivian as a whole. The same is true of the east-adjacent upper reaches of the River Pyrkanaivaam, where the Valanginian-related rocks also contain *Inoceramus* fragments.

Reliable finds of Valanginian fauna were found only to the south-east, in the central part of the Koryak geosyncline (Fig. 39). In the upper reaches of the River Velikaya there occur fauna of the Late Valanginian and Hauterivian. To the southeast (the right-bank region of the Rivers Koiverelan and Velikaya), the most ancient fauna are of the middle-Valanginian. Fauna of the entire Valanginian, including the early Valanginian, can be found farther to the east in the southeastern part of the geosyncline, in the zone of development of the Chirynai sedimentary series and Chirynai (Yagel) melange. Still farther to the south, in the upper reaches of the Rivers Chirynai, Nauchirynai and others, along with the blocks and tectonic flakes of granitoides and Paleozoic sedimentary rocks which, obviously, testify to the rather close disposition of the Khatyrka island arc, there appear fauna of the Berriasian and Late Jurassic (the Volgean age; Fig. 39).

These data presented above suggest that the age of the base of the sedimentary cover in the Koryak geosyncline becomes increasingly older from the northwest to the southeast, i.e. the basement developed unilaterally from the northwest. According to these data, the active eduction opening of the geosyncline occurred from the later half of the Late Jurassic and stopped in the Barremian or Aptian. The corresponding time-span can be estimated at 25—30 my. With the most probable rate of the eduction process of the order of several (say, 3—4) cm a year, the above period would be long enough for the opening of the geosynclinal basin wider than 1000 km, i.e. wider than the Sea of Japan, for example. The effect of the subsequent folding made the width of its corresponding area 250 km and less.

Complications in the opening of the geosyncline: Velikorechensky trough

It is unlikely that any geosyncline, the size of which was comparable to that of recent seas, would have been able to develop with no additional structural complications. This, no doubt, applies equally to the Koryak geosyncline. In its central part, in the basin of the River Velikaya. there is the Velikorechensky trough, represented by a band of width 50-60 km and northeastern strike comprising conglomerate-sandstone-aleurite partly Aptian—Albian and mainly Cenomanian—Turonian deposits, more than 2.5 km thick. They form linear, folded structures, also of northeastern strike (Fig. 39), which are readily visible in the aerial photographs. The trough has distinct boundaries: in the northwest this is the Berezovsky Fault — one of the largest in the Koryak fold area; in the southeast, the margin is defined by the Tamvatvaam zone of faults. The above-described territory of the Koryak geosyncline, with its basement of sedimentary cover of the Late Jurassic—Neocomian age is cut by the Velikorechensky trough into two parts; to the southwest of the River Velikaya basin the trough probably has no continuation.

The Velikorechensky trough can be supposed to have formed by disruption of the geosyncline basement which had existed earlier. The rupture was probably accompanied by the emergence of a new eduction zone which had developed inside the geosyncline at the location of the present is the Berezovsky Fault. This can be surmised from the elevation in the Albian—Turonian of the northwestern wing of the Fault: as a result, there are no deposits of the Aptian, Albian, Cenomanian and Turonian to the northwest of the Berezovsky Fault in a band 25-30 km wide, while they occur to the southeast in the Velikorechensky trough and also farther to the northwest in the zone of the Talovka-Main Rise. The Campanian deposits, thus, in this band immediately overly the Neocomian ones (in the upper reaches of the Rivers Mukarylyan and Berezovayas).

The zone of the Berezovsky Fault proper was created mainly by dislocations along fractures that dip to the northwest, which can be deduced by the distribution of the northwest-inclined tectonic lenses of various composition, including ultrabasic and micromelange rocks in the southern part of the Rarytkin Ridge; they are traced also as an interrupted chain along the zone of the Berezovsky Fault to the southwest. The oldest deposits in the Velikorechensky trough are associated with its southeastern boundary (the Tamvatnei Suite of the Aptian-Albian) which may indicate its eductional asymmetry. According to the viewpoint proposed, a new eduction (paleoseismofocal) zone originated in the Aptian-Albian in the area of the Berezovsky Fault along the line of thrust-apart rupture within the Neocomian eduction crust. The hanging wing of this zone, as always in active structures of this kind, developed into an isostatically unbalanced rise in paleorelief, which in the Turonian, when the Velikorechensky trough opening stopped, disappeared due to compensatory subsidence.

The beginning of the opening of the Velikorechensky trough coincided with the ending of eduction activity in the Talovka-Main paleotrench whose burial began in the Barremian—Aptian. Thus, the Aptian—Albian witnessed an exchange of eduction zones with a transference of the eduction out-thrust line to the southeast in the direction of general outflow of depth material, as a result of which, the direction of out-thrusting remained the same. From this follows that this was not a migration of the same zone, but a replacement of one zone by another which emerged in another place.

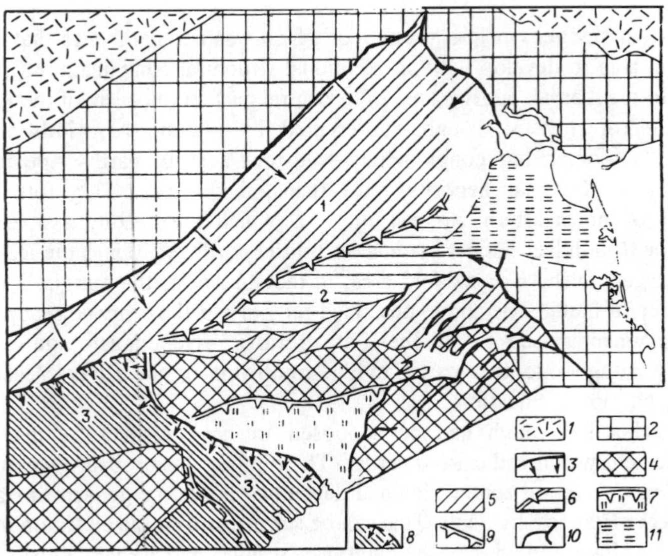

Fig.40. Sketch of development of Koryak geosyncline. 1—Okhotsk-Chukotka volcanic belt; 2—continental slope of Jurassic and Neocomian; 3—line of eduction out-thrusting in Jurassic and Neocomian (Talovka-Main zone and its continuation); 4—blocks of Kamchatka-Hatyrka island arc; 5—eduction crust of Late Jurassic and Neocomian age; 6—line of eduction out-thrusting which emerged in Aptian—Albian (Berezovsky Fault zone) and eduction crust of Aptian-Turonian age; 7—line of eduction out-thrusting which emerged in Coniacian and eduction crust of Coniacian—Early Santonian age, 8—line of eduction out-thrusting which emerged in Santonian and eduction crust of Santonian—Eocene age; 9—possible line of eduction out-thrusting of pre-Maastricht age; 10—system of large overthrusts in eastern Koryak highland; 11—area of Late Cenozoic tensile destruction. Age of the earliest folding: 1—Albian-Turonian, 2—Campanian, 3—Oligocene-Miocene.

Migration of geosynclinal basins and areas of folding after the Early Cretaceous

After the opening of the Velikorechensky trough, the Koryak geosyncline seems to have started a new stage of its development: its eduction basement, which formed in the Late Jurassic—Neocomian and gained considerable width, began disrupting longitudinally (Figs 40 and 41). After the Velikorechensky trough formed, more troughs and basins with the oceanic basement emerged to the south and southeast of that trough. These new troughs and basins successively displaced to the southeast, with the accompanying turn of their strikes from the northeastern to the east-north-eastern and sublatitudinal directions, the latter, closely similar to the trend of recent structures of the northwestern Pacific surroundings. We will not dwell here on the details of their development since the evidence available does not, yet, allow us to trace the changes in the age of the base of the sediments in each of these newly opened basins, i.e. to obtain direct evidence of eduction or any different method of development: this is hindered by complex dislocations in these basins and their relatively small sizes. However, general migration of the new-opening geosynclinal basis to the southeast (which has long been known

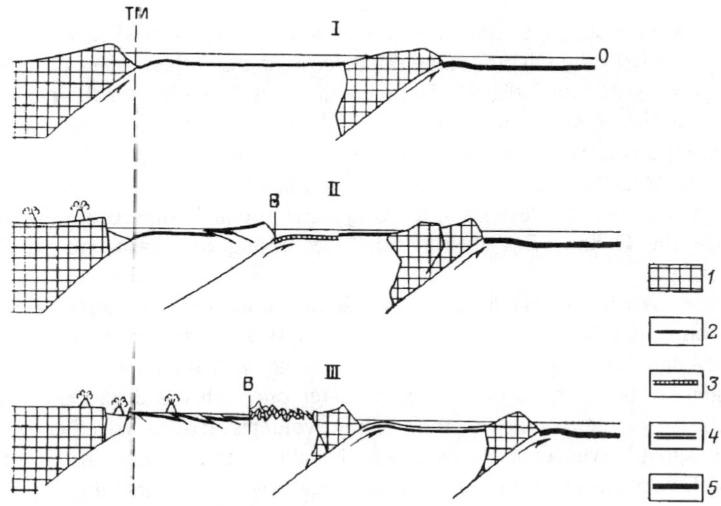

Fig. 41. Opening of Koryak geosyncline. I—Later Jurassic—Neocomian, II—Aptian-Turonian, III—Santonian-Eocene. 1—lithosphere of continental margin and foundation of Kamchatka-Hatyrka island arc; 2—eduction crust of Jurassic-Neocomian age; 3—same, of Aptian-Turonian age; 4—same, of Santonian-Eocene age; 5—eduction crust of the open ocean's margin. Arrows show direction of eductional out-thrusting of depth matter. Position of Talovka-Main trench (TM) is designated with vertical dotted line, B—Berezovsky Fault.

and described by a number of researchers) confirms the general eductional character of development throughout the entire Koryak geosynclinal province.

Formation of new basins was accompanied by the bringing out onto the Earth's surface of new tectonized basic-ultrabasic basement, whose outcrops are of especially wide distribution in the youngest southeastern part of the Koryak geosyncline. When one moves from the northwest southeastward, this basement becomes overlain with increasingly younger sediments. The age of the sediments found together with the basement's outcrops, as well as the cited dating of the supposedly volcanic basement of the sedimentary and of the supposed `serpentine intrusions', essentially indicate the time when the geosynclinal basin formed. To the south of the above-described area where the eduction basement of the Upper Jurassic—Neocomian age is distributed, the basement's age gets successively younger, based on the previously cited information, from the Santonian—Coniacian to the Maastrichtian, southeastward. After the Velikorechensky trough was filled and ceased to open in the Turonian, to the south of it there formed, mainly in the Santonian and Campanian, the complex Enchyvayam-Ukelayat basin which seems to have consisted of several different-age troughs. It seems to be most probable that this trough of northeast and in its eastern part sublatitudinal strike cuts the previously single Olyutor-Khatyrka chain of detached island arc pieces into two parts: the northern, of the Khatyrka blocks, and the southern, which I call the Pakhacha Block (Figs 36 and 37). The latter has no outcrops of the pre-Mesozoic basement onto the surface, but is characterized by development of gently dipping, mainly, Paleogene and Neogene deposits which are mostly volcanic; another feature of it is that it has distinct borders of the block enveloped with folding structures of the Paleogene and Upper Cretaceous geosynclinal sediments. These peculiarities of the Pakhacha Block suggest the existence of an ancient rigid foundation within its borders and hidden at depth. To the south and the east of the Pakhacha Block and of the Enchyvayam-Ukelayat paleobasin there was the Olutor paleobasin, even younger than the former, with most ancient geosynclinal deposits of Maastrichtian age.

During the Maastrichtian, south-east of the previous Enchyvayam-Ukelayat basin and to the south where it crosses the Olyutor-Khatyrka paleo-island arc, there opened the South-Koryak basin which had existed until the late Eocene and which had adjoined the persisting southern part of the paleo-island arc on its north-west. Following the origin of this basin, its preceding Enchyvayam-Ukelayat basin, similar to the previous events, underwent folding and ceased to exist after the Maastrichtian. Sediments, which filled the South-Koryak basin, underwent folding at the end of the Eocene or in the Oligocene, probably due to opening of the Lower-Pakhacha (in the Olyutor Bay district) and the Bering Sea basins in the Neogene and Quaternary.

So, we can observe the regular succession discussed above. The Jurassic-Early Cretaceous folding obviously developed in the rear of the eduction zone responding to the Talovka-Main Rise; we considered this folding to be a result of the horizontal compression forces developing in the hanging wing of the eduction zones. Later on, with the displacement of the acting eduction zones to the southeast, the folding volumes moved together with them and successively spread over their northwestern hanging wings. The opening of the Aptian-Turonian Velikorechensky trough was accompanied by pre-Albian folding (in the Talovka-Main Rise by pre-late Albian) which is manifested widely in the lower reaches of the River Vayega and neighbouring territory; to the northwest of the northwestern restriction of that trough there was probably an eduction region accompanied by a seismofocal zone. The sediments of the Velikorechensky trough underwent folding in Campanian time as in it are observed remains of the continental Maastrichtian deposits overlying unconformably those of the Cenomanian—Turonian; and within the Aptian—Turonian uplift to the northwest of the Velikorechensky trough there accumulated thick sea sediments of the Campanian. Thus, in the Campanian the whole zone of the Velikorechensky trough underwent the folding. The Santonian and Campanian are times of displacement of the geosynclinal opening and eductional out-thrusting to the south into the area of the Enchyvayam-Ukelayat paleobasin; hence, folding within the Velikorechensky trough again took place in the hanging wing of the new eduction zone. In the northern margin of the Enchyvayam-Ukelayat paleobasin, the folding seems to be of pre-Maastrichtian age, while on the whole within this basin the folding ended in the Paleogene. It can be considered to be connected with the eduction opening and existence of the Neogene paleoseismofocal zone northwestern boundary of the recent Bering Sea (which is evidenced by the buried trench revealed seismologically — see Shmidt, 1980). The succession of the paleoseismofocal zone emergence, the opening paleobasins and the developing of fold deformations are shown in Fig 42 .

It should be noted that a close pattern of displacement of different-age paleoseismofocal zones from north-west to south-east was published by several authors—adherents of plate tectonics—after the first edition of this book (Stavsky, *et al.*, 1988; Kononov, 1989) and without going into geological detail.

From the development of the Koryak geosynclinal system one can point out the features which agree with the eduction model of development of paleobasins: 1) successive opening of the basins with a melange foundation in the direction off the paleocontinent, i.e. in conformity with the general direction of eduction mantle outflow; 2) a general increase in age of the base of sediments filling the largest and most studied Koryak paleobasin in the direction farther from the Talovka-Main paleotrench; 3) repetitive folding deformation of the foundation of the paleobasins and their overlying sediments after a new paleoseismofocal zone emerged to the south-east of them.To the northwestern wing of this zone, these basins now become closed, while the opening of new basins without any compressive deformation occurs within the laying wings of new seismofocal zones. The consistent crumpling of the foundation of each paleobasin shows an absence of its rigidity, which is required by the subduction model.

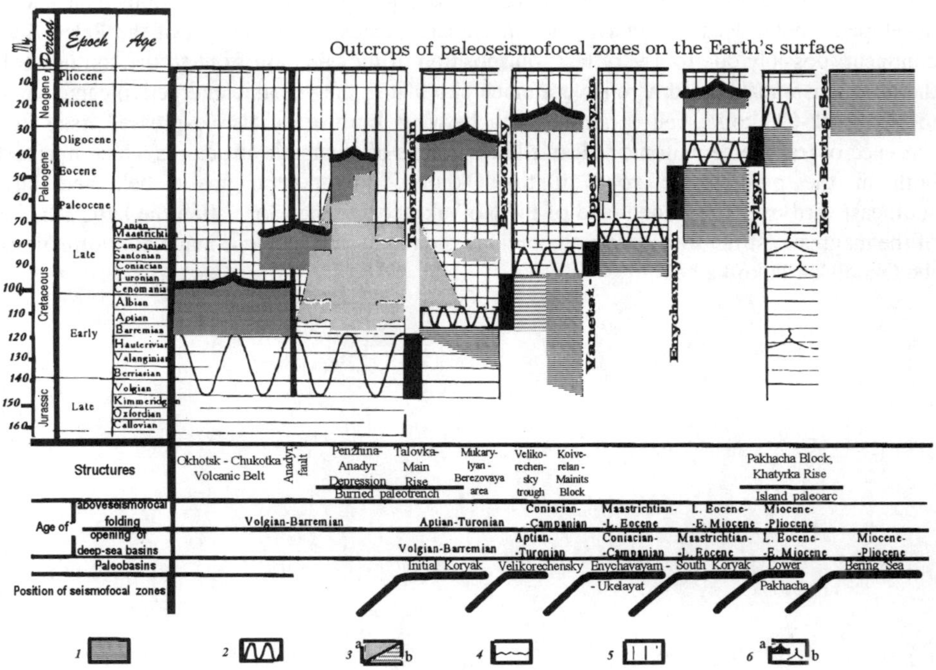

Fig. 42. Diagram of geological development of northeastern continental margins of Paleo-Asia during middle-late Mesozoic and Cenozoic. Along horizontal axis, a schematic geological section across recent Koryak fold area is plotted; on vertical axis, geological time scale. Narrow open columns correspond to sites of outcropping of different-age paleoseismofocal zones on the Earth's surface; time of their existence is shown by black. Signs of geological processes: 1—geosyncline sedimentation, 2—folding, 3—sedimentary filling of Talovka-Main paleotrench and of some minor depressions (a), filling of Velikorechensky trough (b), 4—stratigraphic unconformities and lost strata on continental slopes, 5—post-folding continental regime, 6—volcanism, post-folding sub-aerial (a), island-arc (b).

Migration of volcanism

Similar to the folding but later by about 40-50 my, moving from the northwest to the southeast was an area of magmatic activity (Fig. 42). Formation of the Okhotsk-Chukotka volcanic belt and its perivolcanic zone occurred mainly during the end of the Early—Late Cretaceous. According to Belyi's data, a surge of volcanism took place at the end of the Albian (Belyi, 1977); it can be connected with the restructuring of the eduction system comprising the transposition of the out-thrusting front from the Talovka-Main zone to the southeast in the Berezovsky zone. The Early—Late Cretaceous volcanic belt had a distinctive structural position and was disposed over the Upper Jurassic—Early Cretaceous Paleo-Asiatic edge-uplift. The Senonian volcanic outflows were already displaced considerably to the southeast with respect to this area, into the territory of the Early Cretaceous paleocontinental slope. In the zone of the Early Cretaceous trench (the Talovka-Main Rise), volcanic activities began in the Eocene; in the region of the Velikorechensky trough it took place in the Paleogene-Early Miocene; over the area of the Enchyvayam paleobasin, in the Miocene;

finally, still farther to the southwest in the Pakhacha Block it was of the Pliocene-Quaternary age.

On the folded basement of the former Koryak geosyncline no continuous volcanic belt developed, unlike what had taken place in the Cretaceous on the margin of the Paleo-Asiatic continent, possibly due to a very fast transposition of the eduction front to the southeast. But the magmatism manifested here is very much like that in the volcanic and perivolcanic parts of the Okhotsk-Chukotka belt. The similarity between both areas also expressed itself in the presence of transverse plutonic-volcanic linear chains of magmatic structures which manifested both in the perivolcanic zone of the Okhotsk-Chukotka magmatic belt and farther southeastward over the folded land of the Koryak geosyncline. So, within the latter, the `jets' of the magmatic structures there may continue the range of transverse magmatic formations of the Okhotsk-Chukotka belt.

Chapter V. Deep-Sea Drilling Oceanward from Marginal Trenches

The deep-sea drilling that has been carried out since 1968 by the research vessel *Glomar Challenger* has yielded much evidence on the structure of the ocean floor. The least meaningful data gathered has been that relating to deep-sea trenches. One of the reasons for this is the great depth of the trenches which often exceed the length of the drill column of the vessel (7 km); but even in relatively shallow trenches accessible to drilling (Ryukyu, Central-American, Hellenic), the drilling was not systematic and the evidence gathered is incomplete in many aspects. It should be noted that the correct determination of tectonic processes in trenches does not depend so much on drilling directly into them as it does on drilling in adjacent localities of the oceanic floor representing their outer swell, and further outwards, over 1-2 thousand kilometres towards the open ocean, i.e. at shallower ocean bottom's depths from 4 to 6 km. It is obvious that eductional or subductional displacement of the oceanic basement leads to changes in age of the oceanic floor oceanward from the trench situated at the foot of the continental or island arc's slope. In the case of eduction one should expect that the age of the oceanic floor gets increasingly older when moving from the trench (Fig. 34), whereas in the case of subduction, on the contrary, the age gets younger. The age of the basement at a specific site in the ocean can be reliably determined from a deep-sea hole if it fully penetrated a sedimentary cover adequately characterized by microfossils (which is most often the case). It should be noted that in most deep-sea oceanic holes, especially those drilled in abyssal oceanic plains, the basement of basic composition has, as a rule, a sharp boundary with an overlying sedimentary cover, and the latter commonly contains no volcanic horizons (except for those containing tufogenic material from distant sources); the same holes do not encounter any sedimentary interbeds in the basement under the base of the sediments. Thus the boundary between the sedimentary cover and the basement, as seen from drilling practice, is in most cases quite distinct, with the exception only of areas in the mid-oceanic ridges and some regions in the western and central Pacific which are characterized by an abundance of volcanic islands (where interbedding of sedimentary and volcanic rocks is common).

A method for tracing changes in the age of the oceanic floor to find out its origin has been widely used in the mid-oceanic ridges and has confirmed their spreading nature. To a much lesser degree this method was employed on the ocean margins. As follows from the above, to answer the question of whether eduction or subduction takes place in trenches (or to verify the subduction hypothesis which is also necessary) the drilling sites must be spaced over profiles transverse to the strike of the trenches at a distance from the trench axes of hundreds to some thousands of kilometres, providing the drilling reaches the sedimentary cover's basement. The fact that profiles of such a type have yet not been executed should be considered a shortcoming of the deep-sea drilling programs. This may be explained by the difficulties encountered in drilling holes at such considerable depths of the ocean; another reason may be that the mission has been aimed mainly at confirming the spreading process, whereas most of the crew were already convinced of the reality of subduction to the point where to verify it seemed to be superfluous. However, the results which might be obtained would of course repay all the efforts: by now it would be possible to establish the existence of either eduction or subduction processes, or their absence.

However, at first, until the mid-1970s, the drilling work near the active ocean margins was quite intense, but later (due to the disappointment of the supporters of plate tectonics in the results obtained?) drilling operations were discontinued. Suffice it to say that out of hundreds of boreholes drilled after 1983 by the new research vessel *JOIDES Resolution,* only three (800-802) can be considered to be directly related to a study of age changes of the floor on the active ocean margins.

The boreholes drilled to date on the oceanic margins are not only much fewer than those in the central areas of the oceans but their disposition from the viewpoint of analysis of movements in the oceanic floor is quite accidental. Nevertheless, over a number of localities, the drilling has yielded obvious evidence for an eductional direction of movement of the oceanic floor near the oceans' active margins. These data are well known but not given due attention. Even quite obvious discrepancy between the drilling results and the subduction hypothesis (see Chapter II—data in the Hellenic Trench) fail to make the supporters of subduction aware of its shortcomings.

If the data obtained from deep-sea drilling in the area of trenches do not testify unquestionably to eduction of the oceanic basement in deep-sea trenches, the only reason for this could be an insufficient number of boreholes suitable for analysis. So it is without question that further investigations on the outer side of the trenches have become a necessity.

As already mentioned, the areas over which the drilling sites have been spaced at different distances from the trench, making arrays suitable for analysis of the age changes of the oceanic floor, are not plentiful. Among them are an array of four sites southeast of the Sunda (Java) Trench (Legs 22 and 26 of *Glomar Challenger*); a greater array of sites disposed over a much more extensive area in the north-eastern quarter of the Pacific (Legs 5, 8, 16, 18 of *Glomar Challenger* together with sites of Legs 63, 64 off the North-American coast); several sites of Legs 6 and 32 between the Kuril Trench and the Shatsky Rise; and 3 sites in front of the Mariana Trench (Leg 129 of *JOIDES Resolution*). In none of these localities were the sites designed with the specific purpose of setting profiles transverse to the trench strike, so, from this viewpoint, their disposition was sporadic. The closest to the correct profile line is made by sites of Legs 6 and 32, but it discontinues too far from the Kuril Trench. The drilling results over these localities are considered below. Many boreholes were drilled at different distances from the Philippine Trench, inside the Philippine Trench and inside the Philippine Sea (Legs 6, 31, 58, 59, 124E); however, their disposition is very inconvenient for tracing variations in the age of the ocean floor.

WHARTON BASIN (NORTH-EAST OF INDIAN OCEAN)

The Indian Ocean is mainly surrounded by continental passive-type borders, with no volcanic belts or seismofocal zones accompanying them. The exception is the north-western Indian Ocean adjacent to the Sunda Trench and Sunda Island Arc. It occupies an isolated position and is separated from the rest of the Indian Ocean, which has clear signs of a spreading origin, in the west by a rectilinear meridional submarine East-Indian Ridge, or the Ninetyeast Ridge, stretching over more than 4000 km, and in the south by the West-Australian Ridge and Naturalist Plateau adjacent to the south-western edge of the Australia. The triangle included between the above-mentioned ridges and the Sunda Trench is of a height of about 3000 km (from the north-east to the south-west) and includes the West-Australian and Cocos Basins divided by a range of Cocos Islands; in Reports of the Deep Sea Drilling Project they both are described together as the Wharton Basin.

Views on origin of Wharton Basin

When applied to the Sunda Trench, the subduction model displays obvious contradictions, as it usually does. According to the main scheme of the plate movements, the oceanic crust which was formed in the Mid-Indian Ridge should descend under the neighbouring plate in the nearest Sunda Trench. As Le Pichon (1968; also Le Pichon *et al.*, 1973) pointed out, subduction within the Sunda Trench (and at the same time under the Himalayas) is of special importance for the plate kinematic scheme because it compensates for the expansion of not one, but of three plates all expanding at the same time. In addition to the nearest Australia-Indian Plate, which expands through spreading in the Mid-Indian Ridge, there is also the Antarctic Plate which expands in the Circum-Antarctic Ridge, and the African Plate which expands in the Mid-Atlantic Ridge. All these should have their `sink' in the Sunda Trench because they have no subduction zones of their own. The width of lithosphere moved in this case under the Himalayas, the Sunda Trench should make over 7000 km (Chudinov, 1976).

The assumption of underthrusting of the oceanic floor under the Sunda Arc clearly displays one of the major logical contradictions of the `new global tectonics'. As is well-known, the adherents of plate tectonics usually assume that the spreading ridges arise at the place of breakup of the continental lithosphere—as an example, the Mid-Atlantic Ridge, which has no present-day subduction zones on its borders, is usually used. Should we consider that if the borders with subduction zones existed they also initially contacted the axial line of the mid-ocean ridges? — i.e., the foundation of the Sunda Arc with its Paleozoic rocks abutted along the line of the Mid-Indian Ridge upon the opposite African coast or some other part of Gondwanaland? If this was so, then why, with the presence of the subduction zone, did not the oceanic floor, produced in spreading, submerge right under the Sunda Arc; instead, it moved the latter away over 4000 km? (The same position is true in the Pacific: with an abundance of subduction zones, the spreading foundation proved incapable of underthrusting in them and moved them away from the spreading axis over 12 000 km). If the Sunda trench had always been at some distance away from the Mid-Indian Ridge, then what was the space between them, and what was the determining factor for the initiation line of the spreading ridge?

Furthermore, the deep-sea drilling, carried out by the *Glomar Challenger* in the Wharton basin in 1972-1974 on Legs 22, 26 and 27 (Fig. 43), established the age of the base of the sedimentary cover which contradicts Le Pichon's scheme (see also Kidd and Davies, 1978). The most ancient, Albian age, was proved to be the sedimentation base in the West Australian Basin at Site 256 which is 1600 km from the Sunda trench; at Site 212, which is northeast of Site 256 and 1300 km away from the trench, the age of the sedimentary cover base is Campanian; at Site 213, 800 km away from the trench (true, in the Cocos Basin, while the former two are in the West Australian Basin), the age is late Paleocene. The order of changing age of the base of the sediment at these three sites relative to the trench corresponds, thus, to eductional out-thrusting of the sea floor, as we must consider that the origin of the latter in the West Australian and Coços Basins, disposed equally relative to the Sunda Trench, should be identical. The eductional order looks distorted at the fourth Site 211, which is 200 km away from the trench axis on the outer slope of the oceanic swell because, at the base of the sedimentary section at this site, nannofossils of the Maastrichtian and Campanian—Maastrichtian age were found above the oceanic basement (*Initial Reports..., 1974a).* One should point out, however, that the clays containing them make up only an 18-m interval and bear traces of deformation; they are also separated from the overlying continuous sedimentary section 400 m in thickness by a large stratigraphic discontinuity, since the age of this section at its base is Pliocene. The above-mentioned Cretaceous nannofossils are poor in their assemblages and have signs of redeposition; the same is true of foraminifera found together with them which contain no plankton, but only benthos forms, some of which are analogous to those typical of shelf sediments; the latter is confirmed by the presence of *Inoceramus* prisms (*Initial Reports ..., 1974a*). These allow a supposition that the true sedimentary section at Site 211 begins only with Pliocene deposits whereas the underlying deposits with Late Cretaceous microfossils make up a displaced body of landslide or tectonic origin. In this case, the section

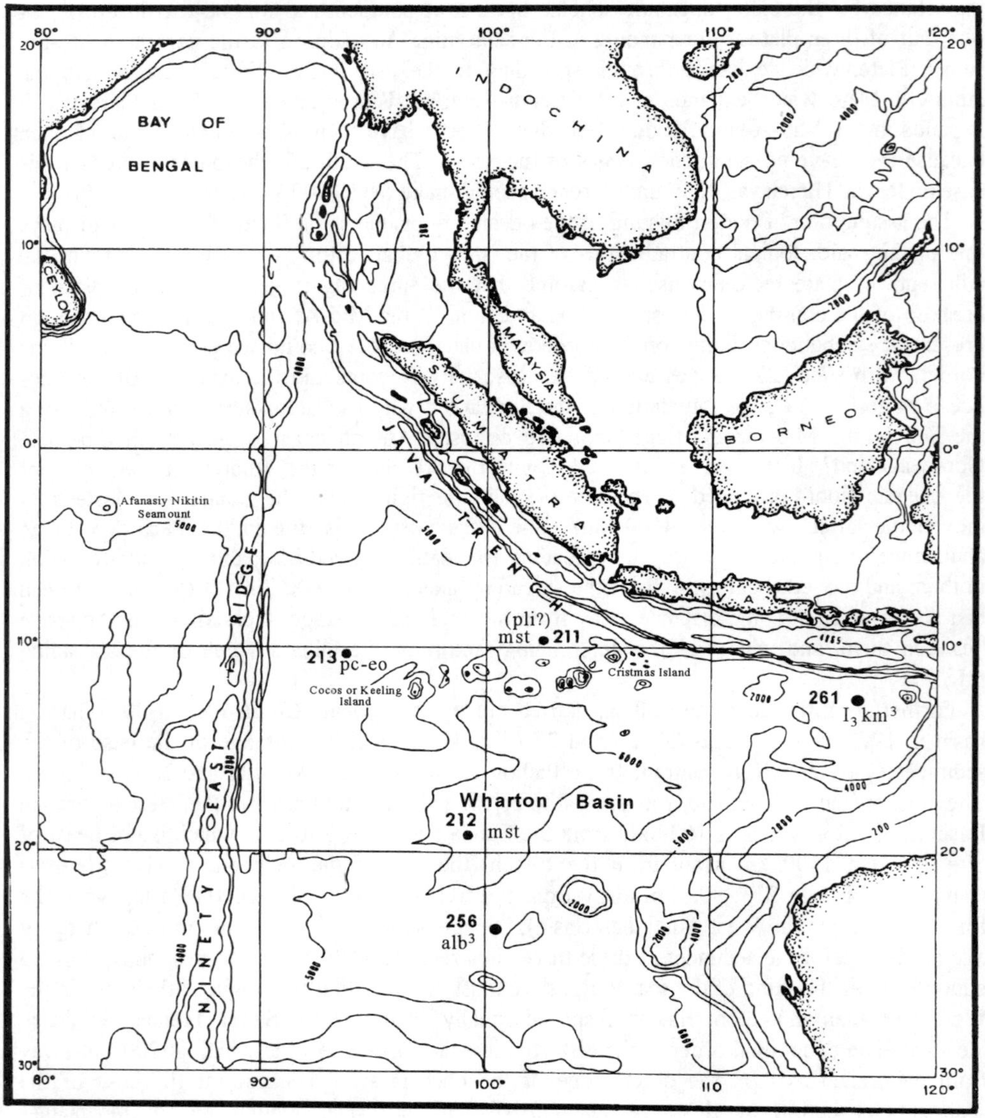

Fig. 43. Location of drill sites in the Wharton Basin in front of the Sunda Trench. (Initial Reports of Deep Sea Drilling Project). Age of sediments at the base of sites' sedimentary sections: alb^3- Upper Albian, mst - Maastrichtian, pc-eo - Paleocene or Lower Eocene, (pli?) - possible Lower Pliocene, I_3 km - Upper Kimmeridgian in the Argo Basin.

of Site 211 continues the eductional succession established from other sites of the Wharton Basin. The rate of eductional displacement away from the trench determined for each interval (Sites 256-212-213-211) turns out to be rather similar at 1—2 cm per year. The mean rate of eductional out-thrusting which is derived from division of the distance from the Sunda Trench to Site 256 by the time which passed after Albian (100—110 my) is close to 1.5 cm per year.

However, even if the age of the base of sedimentary deposits at Site 211 is assumed to be Campanian-Maastrichtian, the subduction underthrusting of the floor in the Sunda Trench according to Le Pichon's scheme is not possible either. The age of the sedimentary cover base at Site 211, which is near the trench, and at Site 212, which is more than 1000 km from the trench in the direction transverse to it, turns out to be similar, whereas at Site 256, which is another 300 km farther, it is much older, Albian. This circumstance made J. Sclater, C. von Borch, and J. Veevers *et al.* (*Initial Reports..., 1974a;* Sclater and Fisher, 1974) to essentially review the scheme of plate movement in the western Indian Ocean. The magnetic anomalies known for the West Australian and Cocos Basins which are of a piecemeal nature and are mainly of sublatitudinal strike, were interpreted as gaining in age from north to south, i.e. in the reverse direction reverse to that of Le Pichon's scheme. However, these authors supposed that in the Wharton Basin, it is not the north-eastern wing of the Mid-Indian Ridge which is located there but the opposite southwestern one, because of the sudden 'jump' of early Eocene of the ridge crest onto a new location. As a result, the western part of the Mid-Indian Ridge, which was until late Paleocene disposed in the northern half of the Indian Ocean and had a west-east trend, was displaced in such a way to its present position, while the eastern part of the ridge was either hidden beneath the Indonesian Archipelago or entirely disappeared in some other way — the authors did not give explicit details. Two variants of explanation, differing little, were proposed and are illustrated in two figures (idem, p.p. 828, 829).

On the whole, such a scheme for development of the Indian Ocean with such a sharp displacement of the mid-oceanic ridge is rather arbitrary. An eductional origin for the Wharton Basin (together with its western continuation — Mesozoic Tethys — which later was folded and destroyed as a result of the northward movement of the Indian Platform, according to R. Ditz's scheme — see the end of Chapter II) seems much more probable.

Sedimentary cover in drilling holes

Let us consider the sections of sites in the Wharton Basin in further detail.

The hole of *Site 211* (Leg 22) was drilled southwest of the Sunda Trench just behind its outer swell. The floor depth is 5528 m, the penetration of the sediments and basement is 447 m, and the thickness of the sedimentary cover is 429 m. From the sea bottom down to the depth of 304 m the section is represented by diatomaceous and radiolarian oozes, clayey aleurites, clays, aleurite and clayey sands, the last of these occurring from the depth of 100 m and deeper and showing signs of turbidite constitution. At all levels, the section contains tufogenic and volcanic matter, mainly of acid composition. In the upper part, the sediments are characterized by Quaternary microfossils, and from a depth of 57 m by those of the Pliocene, mainly by radiolarians. The cores from the interval of 95-190 m displayed only radiolarians of mid- or late Miocene which are considered to be redeposited, since down below, at the interval of 228-304 m, the cores contained foraminifera not older than Pliocene (sampling of cores in this and most other holes was carried out with gaps). Farther down, some cores of brown clays and aleurite sands with no faunal remains were raised from depths of 342-351 m; also brown clay, not characterized faunally, was raised from the depth of 398 m. Three metres deeper there began a 10-m interval described as diabase sill (also referred to in some places as andesite diabase). The core was taken from the lower part of the interval of 409-411.6 m. In its composition the fine-grained 20-cm layer was described on the lower contact of the sill and was considered to be the sill's chilled margin. According to the description, this layer is especially weathered; however, this is not typical of chilled zones of basalts and diabases which are normally most resistant to weathering.

In three cores from the following 18 m (0.8 m right below sill, 1.15 m from the depth of 419 m and 0.65 m from the depth of 428 m) directly above the oceanic basement, variegated red, brown and grey clays and clayey oozes were raised which contained up to 50-80% of remains of calcareous nannofossils. The middle core contained interbeds of several millimetres of ash tuffs with fragments of greatly altered volcanic glass and a high content of hydrous ferric oxides and mica; its lower part has sections of microfolding, with wavy contacts of many beds and foreign inclusions. The nature of the foraminifera and the prisms of *Inoceramus*, as was mentioned above, indicates the probability of an offshore origin for these sediments. The remains of nannofossils (coccoliths) are corroded and destroyed, especially so in densely colored layers and detectable only in a small part of the index species. The remains discovered in the cores are dated as early Maastrichtian—late Campanian.

The hole of *Site 212* (Leg 22) was drilled 1200 km to the south-southwest of Site 211, 1400 km to the southwest from the Sunda Trench. The depth of the floor is 6420 m, the depth of the hole is 521 m, the thickness of the sedimentary cover is 516 m. Down to the depth of 288 m there were revealed nannofossil oozes of yellowish and brown tints with interbeds of brown clays near the surface and with the chalk horizon at the base. The age of these deposits corresponds to the Pliocene at the top and to the beginning of the mid-Miocene at the base. Then follows an interbedding of zeolite brown clays about 30 m thick without identifiable organic remains, and about 80 m of chalk of Mid-Eocene age. From the depth of 400 m down there is again a 30-m horizon of brown zeolite clays and then 50 m of nannofossil chalk but already of Late Cretaceous, Maastrichtian, age. The chalk is again underlain by a horizon of dump zeolite clays, 35 m thick, similar to the previous ones and overlapping bedrocks diagnosed as spilitic metabasalts; they were drilled for 4.5 m.

The hole in *Site 213* (Leg 22) was drilled at the depth of 5609 m, the penetration is 172.5 m, the thickness of sedimentary cover is 154 m, 18.5 m being drilled on the basement rocks. The top 70 m are represented by radiolarian-diatomaceous oozes of reddish-yellow coloring. The clayey fraction of the oozes constitutes just a few percent.

Over the whole section one can find volcanic ash of acid composition making up as much as 25% of the ooze, and pumice fragments were observed in its top and middle parts; the age of the sediments was from Quaternary to late Miocene. The following 70 m are made up by brown zeolite clays. The content of zeolites, mainly phillipsite, in clays varies from 1 to 20%. In the lower half of this interval, ferric oxides and manganese micronodules, whose content grows downwards up to 15%, appear in the clay. At the top, the presence of liparite volcanic ash is noticed. There are no microfossils in the zeolite clays but they appear lower in the nannofossil oozes, 12 m thick, whose top contains early Eocene, and the lower part mixed early Eocene-late Pliocene foraminifera and nannofossils. Then the succession moves to basal barren ferro-manganese sediments. Three cores were obtained from the basement rocks (from the depth of 183-156 m, 160.5-163 m and 164-166 m). According to the description, they contain eleven flows of pillow lavas represented by basalts and separated by `chilled zones' of palagonitized glass.

Site 256 (Leg 26) is 1600 km away from the Sunda Trench. The bottom depth is 5361 m, the penetration is 270 m, the thickness of the sedimentary cover is 248 m. The whole section is represented by thoroughly homogeneous clays with admixture of aleurite material and characterized at the top by Quaternary and Pliocene microfossils and, from 180 m down (after a 50-m interval where no core was taken), by Late Cretaceous ones. In the interval from 237 to 248 m (the section's base), the sediments contain late Albian fossils. The Cretaceous sediments differ from the overlying ones in greater strength which was reflected in the character of deformations of the core. A thickness of 19 m was penetrated in underlying basalts which are, according to the description, greatly altered in some parts and as-formed in others.

The base of the sedimentary cover of all sites in the Wharton Basin reveal signs of underwater currents and, possibly, a proximity of the island slope greater than at present. In Site 211 the base of the Quaternary—Pliocene (if it alone is considered to be autochthonous) is represented by aleuritic sands and clays with some fragments of calcareous aleurites. In the last 50 m the core catcher (no core was obtained) seized essentially ferrous material with up to 90% of iron oxides and admixture of clay and volcanic ash. Such material may be analogous in its origin to essentially ferrous sediments settling from the gels, whose present deposition is conditioned by carrying out of mineralized solfataric waters of terresrial volcanoes into the offshore zone. This was observed on an underwater slope of the Kuril Islands by Zelenov (1972); the same author cites a description of such phenomena in the Indonesian Archipelago. Site 213 reveals at the base of the early Eocene—late Paleocene section a layer of analogous material of about 5 m in thickness with content of not only iron oxides (up to 80-85% of the sediment) but also oxides of manganese. In Site 212, the base of the section is represented by nannofossil oozes containing up to 30% of transferred and partly destroyed foraminifera of Campanian-Maastrichtian.

Report on Leg 22 (*Initial Reports..., 1974a*) discussed the cause for the presence of considerable amounts of carbonate material at the base of the hole's sections, below the level of carbonate compensation. It was suggested that the cause was transportation and fast burial of material deposited at a higher level, or that the compensation level in the past was deeper than its present position (p. 580). Since the former supposition may be thought more plausible, from a number of considerations (in particular, of above-mentioned signs of rewashing of some components), its substantiation should take into account the possible initial proximity of all localities of the sea floor in the case of eduction from out of the slope of the Sunda Island Arc. This relieves the main difficulties in explaining the transportation of carbonate material by underwater currents, which implies considerable present remoteness of the oceanic floor under consideration from the land elements of relief.

Composition of oceanic basement in Wharton Basin

Special consideration should be given to the character of the basement of the Wharton Basin that has been revealed by the drilling. As has been mentioned above, from the eduction viewpoint, the oceanic basement adjacent to the trenches being of another origin should differ from the spreading basement. According to the data available, on the whole, the latter is of a volcanic character, while the eductional foundation is dynamometamorphic. It consists of a mixture of rocks which were not formed *in situ*, but were tectonically transported during the eduction process from the side of the continents or island arc, and hence many of them were brought from great depths. Thus, we are dealing with a tectonic mixture, or melange. The question of the latter's presence in the basement of the recent oceans arises, on the one hand, as a consequence of a logical development of the eduction model, but, on the other, this question ensues from actual parallels between the basement in the present oceans and in the ancient oceans transformed by folding. It is known that within the latter, basic-ultrabasic melange is widely spread and is usually overlain by ancient sea sediments which testifies to its belonging to the oceanic basement. In Chapter IV above the melange basement of the Mesozoic oceanic provinces in the northwestern Russia was described, and cases of it being overlapped with Mesozoic sediments were also reported.

As it was shown (and is well known), such a basement is partly represented by large-block melange which contains erratic bulks of rocks of a hanging wing of eduction zones with sizes up to several kilometres (for example in the Mukarylyan-Berezovayas region: if one imagines that within it there would be assigned randomly positioned boreholes, then approximately every one out of four would hit on a large erratic piece embraced with the predominant basic-ultrabasic mass; for this reason the question was posed above of a possible presence of a detached piece of deformed Late Cretaceous deposits at the base of the sedimentary section of Site 211 which is not the continuation of the latter). Further, widely distributed is melange with blocks, fragments and rolled balls of ultrabasic rocks with sizes in metres and decimetres, which partly preserve inside them their initial textures, included in serpentinized mass: this melange also contains fragments of basic rocks, of metamorphic rocks from mica-chlorite schists to eclogites, as well as of sedimentary rocks from the ancient surroundings of the basin. Finally, widely distributed is micromelange of serpentine and basic composition which, due to dynamometamorphic transformations, has completely lost the textures of the primary rocks and acquired secondary mylonite textures. This material, as with the other varieties of melange, is distributed over wide areas, and, with its predominant basic composition, is easily taken for deeply altered volcanic rocks.

The chemical composition of most micromelange corresponding to basalts is unlikely to be accidental. Apart from the fact that the material of the gabbro (geophysical basalt) layer of the Earth's crust is delivered eductionally onto the Earth's surface, fusion of the basaltic melt both from the rocks of the Earth's crust and from the elevating ultrabasic material of the upper and lower mantle undoubtedly takes place in the eductional out-thrusting zone. In many cases, this melt seems to have time to solidify before it arrives onto the Earth's surface; therefore, the presence of basic rocks in the melange composition, showing signs of crystallization from magmatic melt, should not be considered in all cases to be evidence of outflow of basalts on the oceanic floor. The signs of the origin of the rocks in melange should be seen in their irregular composition, in composition changes over short distances, and in various traces of dynamic transformations and presence of obvious metamorphic rocks.

The evidence of distribution of melange formations in the trench-adjacent localities may be obtained, as a first approximation, from statistics of distribution of differently altered basaltic rocks in the composition of the oceanic basement. Thus, according to Dmitriev *et al.* (1976), the basement of the mid-oceanic ridges contains fresh basalts in the amount of 50.0% of all outcrops and weathered ones make 5.9%, in the trenches 28.9 and 33.4% respectively; the rest of the rocks are made up by ultrabasites, gabbro and amphibolites. The question of the distribution, in particular, of altered melange varieties in the composition of the oceanic foundation should be considered crucial in studying the oceans' origin.

Let us further consider the data on the basement revealed by drilling in the Wharton Basin. It is characterized by considerable alteration and inconsistent composition of melanocratic rocks of which it is constituted. These rocks are diagnosed as paleotypic basalts (spilitized, amphibolous, and others) and partly as amphibolites; many rocks in the Koryak and Oloi-Anyui geosynclines, which were the Mesozoic marginal seas in northeastern Russia are similarly described.

The diabasic sill of *Site 211,* about 10 m thick, which separates the base of the Pliocene section and an 18-m interval with fossils of the Late Cretaceous, as well as this interval as a whole, can be thought of as a block having lost links with its native massif within a landslide or as an eductional detached piece. The rocks are considerably altered; their characteristics given in the description of the hole and in a special paper by R. Hekinian (*Initial Reports, 1974a,* p. 413-448) are somewhat contradictory. The most widespread minerals are plagioclase (on an average 40%) and clinopyroxene (up to 25%). The plagioclase, andesine No 40-50, occurs in grains of up 2-3 mm in size with polysynthetic twins, with zoning, and strongly altered; many of them are completely replaced by secondary products. Pyroxene (reddish titanaugite) is in finer grains. The texture is described as diabasic, ophitic and subophitic, but hypidiomorphizm is mentioned for both plagioclase and pyroxene. Furthermore, the rock composition contains phyllosilicates (21%) represented by serpentine and alternating layers of vermiculite-chlorite composition, up to 5% of magnetite and hematite, often accompanied with biotite flakes (7%) and needle amphibole (less than 3%). The admixture of calcite, zeolite and pyrite is ubiquitous. The coloring of the rock is black-green, and there are calcite and pyrite veins. No microphotographs are given in the Report. As mentioned above, the sill ends up at its base with a 20 cm weathered zone where the space between fine plagioclase laths is filled with palagonite, calcite, iron oxides and pyroxene.

In spite of much plagioclase in the rock and its relatively low basicity (the sill, as mentioned above, was described not only as diabase but also as andesite), chemical analyses of four samples established very poor content of SiO_2 — from 46 to 41%, — and unusually high content of K_2O — 1.87-2.81%, with Na_2O more than 3%, MgO 3. 99 — 9.63, CaO 5.70 — 11.37% and TiO_2 more than 2%. The composition of the rocks is, thus, anomalous and very unstable. The loss from ignition amounts to 6.72% and the content of CO_2 is up to 4.90%. The authors of the Report explain the peculiarities of the composition as resulting from alteration of the rocks, but they do not consider dynamometamorphic changes which, according to the descriptions, seems to be possible.

The basaltic basement in Site 211 was penetrated to 19 m, which yielded about 9 m of core. The rocks are described as altered ('weathered') amphibole basalts at the top and amphibolites at the bottom. Amphibole basalts are characterized as consisting of two varieties interconnected with mutual transitions: more coarse-grained phanerite and fine-grained aphanite, both being weathered. Where one rock changes over to another, one can observe especially intense alteration; within a distance of some 8 m, such changes may occur twelve times. The terms phanerite (i.e. porphyritic) and coarse-grained are used to describe those varieties whose plagioclase phenocrysts are 0.3—1.6 mm in length, i.e. their coarse granularity is very relative. The total content of plagioclase, including largest grains related to phenocrysts as well as fine laths of the basic mass, makes up about 50% of the rock, and some of this plagioclase occurs as altered basic andesine or secondary albite. The space between the plagioclase is filled with dark material determined as altered glass and palagonite. There are also pyroxene and olivine (as fine grains of tenth fractions of a millimetre) completely replaced by secondary products. The amphibole content is not mentioned but, judging from the name of the rock, it should be considerable; it is little changed and forms prisms of yellowish-green coloring. The rock also contains parallel-oriented needle segregations of brown mica of hundredth fractions of a millimetre in length. Both amphibole and mica are often associated with calcite and penetrated with it.

Amphibolites make up about one third of two cores obtained from the basement, occupying two intervals of several decimetres each. It is believed that they make up flows limited by chilled zones and there are four such flows. The rocks consist of about 80% of amphibole (common and brown basaltic hornblende) with grains, according to the description, of up to 0.4 mm. However, the illustrated photomicrographs show radial intergrowths of amphibole with the length of individual crystals, according to the accompanying scales, of from several mm to 1-5 cm (*idem:* p. 441). In some cases, basaltic-hornblende is overgrown with greenish-blue amphibole (of a glaucophane group?). Furthermore, the amphibole rock has in it several percent of olivine replaced mostly with calcite and iddingsite, and mica without more precise definition. The rest of the rock — up to 20% — is represented by calcite and groundmass which consists of a glass, opaque mesostasis and unidentifiable cryptocrystalline aggregates.

Of interest is the chemical composition of the amphibolites. The content of SiO_2 varies in them from 38.01 to 42.56 %, i.e., corresponds to ultrabasic rocks rather than to basalts. Even if one excludes very high losses from ignition of combustible and volatile components, varying from 6.13 to 11. 43%, the content of SiO_2 will make up only 45—47%. At the same time, the rocks display very high content of K_2O and Na_2O of 1.06—2.01% and 2.78—4.00% respectively; the content of MgO is 4.32—6.29 %, i.e. lower than in some samples of andesite diabases of the overlying supposed sill. In their composition, the amphibolites as well as rocks of the `sill' are more like a mixture of heterogeneous matter.

The authors of the Leg 22 Report believe that the amphibolites described were primary effusions formed from a water-oversaturated basalt melt. One of the arguments favouring this opinion is the presence of glass in the amphibolites. Doubts as to whether identification of all non-transparent masses present in the composition of the basement rocks as a glass is correct were expressed above. It is more likely that the rocks which are classed in the Report as amphibole-bearing basalts, along with the amphibolites proper, should be called, according to the descriptions and photomicrographs, amphibolites of metasomatic origin, implying that the radial intergrowths in them should not be considered to be the remains of variolitic texture. These are most probably nematoblastic and fibroblastic metamorphic textures typical of such rocks, the more so since their amphibole constituents are from several millimetres up to several centimetres long (*Initial Reports..., 1974a*, photomicrograph on p. 441).

Metabasalts from *Site 212* are essentially altered rocks which consist of fine-grained, up to cryptocrystalline, masses of altered plagioclase, amounting to 47 % of the rock, and mesostasis consisting of brown glass, ferric oxides, chlorite aggregate and up to 39% of smectite. The rest of the rock is represented by calcite, epidote, palagonite and supposed volcanic glass. Mostly albitized, altered plagioclase forms microlites in some (though not in all) of the metabasalt varieties, and up to 10% of microphenocrysts; plagioclase phenocrysts are more frequent. The sizes of microphenocrysts indicated are 0.06—1.0 mm and that of microlites is 0.3 mm (*Initial 0Reports..., 1974a*, p. 419); but in some varieties, laths of plagioclase replaced with chlorite and with albite rims amount to 10 cm in length (*idem*, p. 442, photomicrograph).

The penetrated 4.5-m sequence of metabasalts is divisible into three parts by gently dipping zones, of up to 10 cm thick, of a dark-green mass considered to be chloritized basalt glass. The localities divided by them are taken for basalt flows. Analogous zones, considered to be glass, are also present in the top and bottom of the interval. An analogy may be drawn between these glassy-like zones and the ancient geosynclinal formations described above; in this and similar cases they could be zones of basic mylonites along gentle dipping displacement planes. This can explain the change in rocks which differ quite markedly, according to chemical analyses, from one part of the drill core to another and an intense alteration of the supposed chilled zones, where loss from ignition amounts to 19%. Such an assumption can also be made because the identified thicknesses from 1-1.5 m to several decimetres are too small for basalt flows, at least those which usually occur. True, these flows are partly treated as parts of pillows in pillow-lavas; however, one cannot find any descriptions of siliceous rocks which often occur in the gaps between the pillows of the ancient pillow-lavas observed on dry land.

In the hole of *Site 213*, about 20 m of oceanic basement was penetrated. In the composition of eleven supposed flows of pillow-lavas identified in 6.25 m of the cores, a description is given for `weathered' and `less weathered' basalts as well as of `chilled zones' consisting of palagonitized glass (or glassy-like material). The rocks are of light-brown and olive-grey colouring and contain many veins and veinlets of calcite and palagonite. In the least weathered varieties of rocks there occur agglomerations of calcic plagioclase phenocrysts of 1.6-4.8 mm in length and 0.6-1.6 mm microphenocrysts of the same plagioclase. Plagioclase is partly replaced by calcite. Present in the matrix are small amounts of fine plagioclase microlites, clinopyroxene, palagonite, serpentine and smectite, as well as olivine crystal completely replaced with iddingsite, serpentine and ferric oxides. Varieties characterized as glassy and those transient to them are much more weathered. They consist of ferro-brown aggregate of smectite, reddish-brown amorphous limonite and `fresh glass'. Two lines below, in the same description, this fresh glass is characterized as completely weathered (*Initial Reports...,1974a,*

p. 422). Plagioclase is the only discernible mineral but, when in large grains (1-6 mm), it is normally replaced with calcite completely.

Basement rocks in Sites 212 and 213 are probably of magmatic origin but may relate to diabases of hypabyssal origin, since the phenocrysts in them amount to several centimetres in length, if one judges from photomicrographs and their accompanying scales (*Initial Reports..., 1974a*, p. 442).

Basalts of *Site 256* are characterized in a number of photomicrographs of thin sections which clearly show hyalporphyric and other textures, leaving no doubts as to the igneous nature of the rocks. In the 19-m section of basalts, several layers (flows) are identified: these are fresh basalts separated by zones of strongly weathered, especially fine-grained, varieties taken for the roofs of the flow; but the tectonic nature of such zones also seems to be possible. Three chemical analyses established a rather homogeneous composition of the basalts characterized by a content of SiO_2 close to 49% with a high content of TiO_2 (2.5%) and iron (*Initial Reports..., 1974b*).

The rocks of the basement penetrated in all the sites of the Wharton Basin contain considerable amounts of fine veins and veinlets which consist mostly of calcite, but also noted are an admixture and individual veinlets of prehnite, palagonite, smectite and phillipsite.

Descriptions of rocks occurring in the Wharton Basin basement draw attention not only to continual references to great alternation of basalts and their constitute mineral grains (which are supposed to be due to halmyrolysis, submarine weathering and hydrothermal processes) but also to a number of features alien to common basalts, including those of spreading ridges and also most known paleotypic varieties. Although described in many cases, the sizes of phenocrysts (among which microphenocrysts always prevail), as well as of plagioclase grains within the matrix are considerably less than normal. The sizes of phenocrysts are mostly estimated at 0.3—1.5 mm, occasionally at up to 2—3 mm (their content is 1—2%); microphenocrysts are of 0.06—1 mm, while microlites are in the range of thousandth fractions of a millimetre to 0.3 mm. The very classification into microphenocrysts and microlites is thus quite arbitrary, as such phenocrysts and microphenocrysts cannot normally be separated from microlites. (The same sizes of mineral grains can be found in the descriptions of altered basalts of the basement of other trenches; for example, Sites 290, 291, 292, 294 and 296 of Leg 31—*Initial Reports..., 1975a* — drilled in front of the Philippine and Ryukyu trenches reveal foundation rocks with plagioclase phenocrysts sized 0.1—0.5 mm, plagioclase and pyroxene microphenocrysts 0.05—0.5 mm and plagioclase of correspondingly smaller sizes in the matrix. This is smaller by an order than are the sizes of grains normally observed in submarine basalt outflows, in particular, in spilites of ancient complexes outcropping onto land).

In some varieties, along with the microlite mass or without it, only fine-grained plagioclase matrix is mentioned. Plagioclase in all its developments is normally altered and in many cases fully replaced with secondary minerals. Mentioned among the latter most often are albite, clayey mass, chlorite, calcite and iron oxides; sericite is rare, while saussurite seems to be totally absent. The second most distributed, after plagioclase, is clinopyroxene in grains of up to 0.5 mm amounting to 30% of the rocks; the third constituent is glass, which needs special attention herein.

Glass makes up interstices between grains and other minerals, individual parts of rocks of less definite shape and strictly defined fragments. Its total content makes, in most cases, 30-40% of the rock, occasionally up to 60% and more. Occurring in cores are some glassy beds up to several tens of centimetres and also irregular glassy segregations. This glass is not recrystallized; it is a macroscopically black-green, often green, microscopically brown non-double-refracting mass, which is diagnosed as glass precisely because of this last property. However, from a study of ancient basalt rocks occurring on land it is well known that the main sign conditioning their paleotypic nature is completely or almost completely recrystallized glass, which is represented by a crystalline, not isotropic, aggregate of secondary minerals which are clearly discernible microscopically. In particular, in spillites (to which many of the rocks revealed by deep-sea bores are related) glass is normally fully replaced by an aggregate of green chlorite, and often a rather large-scale one. The rocks of the basement in the Wharton Basin and other oceanic areas near trenches sometimes display chlorite-containing glass (or another rock that is so called), but it is almost never replaced with chlorite completely. At the same time, most of such glass is, according to the descriptions, replaced with clayey substance, iron oxides, palagonite, or smectite, which are almost as isotropic as glass proper. There is, thus, an obvious discrepancy between the well-known signs of paleotypic basic volcanic rocks found on land and their alteration, when their extrusive origin is quite certain, and descriptions of rocks obtained from holes in the vicinity of trenches.

It seems quite probable that material taken for glassy paleobasalts is partly mylonite mass of basic composition relating to micromelange. Such mass indeed (as well as mylonites of more acid rocks) often macroscopically resembles a glass.

One could also mention other unusual properties of paleotypic rocks of the basement of the Wharton Basin. It has already been noted that among them, besides varieties related to altered basalts, are distributed amphibolites and petrographic varieties transitional to them. So, amphibolites, passing over into altered basalts and bound with glassy, supposedly chilled, zones, are found in Site 211. They are taken for flows of specific water-oversaturated basalt magma; however, the existence of such magma which turns, when solidified, into rock consisting mainly (80%) of hornblende is implausible. As one of the arguments for the volcanic origin of the amphibolites, the variolitic arrangement of plagioclase laths is most likely an orbicular structure characteristic of metamorphic rocks and formed during metasomatic processes. The reports on deep-sea drilling contain, unfortunately, too few photomicrographs of the basement rocks; there are four of them (*Initial Reports..., 1974a*) for rocks of Sites 211, 212 and 213. Two photomicrographs relating to Site 211 (of amphibolites and amphibolite-containing basalts) show, as mentioned above, large radiating arrangements of crystals, several centimetres in size, which are typical of amphibolites of metamorphic origin.

Chemical analyses of rocks from the Wharton Basin basement testify to considerable deviations in their composition from normative compositions of basalts and, which merits special emphasis, to wide and irregular variations in content of rock-forming oxides. Thus, averaging several sample analyses, we find the content of SiO_2 varies in the range of 46—51% (in some particular samples it drops to 42—39%), CaO varies from 6 to 13%, MgO from 5 to 9%, K_2O from 0.25 to 3.26%, which is unusual for basalts, Na_2O from 1.42 to 3.72; the loss from ignition make several percent. When the changes in rocks are characterized, the authors of descriptions (*Initial Reports..., 1974a;* Thompson *et al.*, 1978 and others) relate them to zeolitic and green-schist chlorite-quartz-pumpellyite grade of metamorphism.

The basement rocks revealed by drilling contain, as a rule, veins of calcite, zeolites and prehnite. This and other features bring them close to the above-described micromelange of geosyncline provinces in the North-East of Russia. Such veins are well seen in photographs of the cores of many sites, in particular, of Site 261, Leg 27 (*Initial Reports..., 1974c*) which is in the southeastern part of the Wharton Basin adjacent to Australia (Argo abyssal plain); here, the most ancient deposits in the Basin, Upper Jurassic (Upper Kimmeridgian) sediments and, underlying them, the Jurassic mafic basement were visible. We will not dwell at length with this description, since close disposition of the site to the northwestern margin of the Australian continent (about 20 km from the foot of the continental slope) makes it probable that the initial position of these deposits was different; they have probably travelled over a considerable distance to the Northeast due to the known movement of the Australian Platform (Chapter III). Turbulent conditions for sedimentation which existed here, beginning from Late Cretaceous, are seen from the borehole section which is represented at its top essentially by clayey olistostromes (*Initial Reports..., 1974c*). The presence of Upper Jurassic sediments in Site 261 testifies, however, to the fact that the Wharton Basin existed probably as long ago as the Late Jurassic.

It should be noted that the absence of extrusions and the presence of a basement represented mainly by a mafic mass that has been ground are not incompatible with an eduction process. Volcanism can develop in eduction basins on the outer oceanic swells of the trenches, but in some cases it may seem to manifest itself in other conditions too. Therefore, overthrust of sediments on altered volcanic rocks does not rule out eduction. Just the same, the presence of individual bodies which no doubt formed from a melt among the altered mass of obscure origin, does not offer any proof for volcanic origin of the whole mass, since in micromelange, according to land observations, there are some rather young vein bodies and dykes. Finally, the eduction base of the basin may include different-size detached pieces of rocks making up the foot and deeper parts of the continental slope, among which one can also find effusions altered to a different degree. Therefore, the extensive distribution of altered mafic rocks, which acquire in places distinct features of metamorphic formations, when their classification into paleotypic volcanic rocks seems to be arbitrary, can be considered as a sign of the eduction origin of the basement in the basin under study. It is oceanic basement of such a type that is distinguishable from that originating from spreading, which is represented normally by slightly altered basalts. The former can be observed in the Wharton Basin as well as over many other localities of oceanic floor opposite deep-sea trenches.

PACIFIC OCEAN

Northeastern part of the Pacific

The earliest legs of the *Glomar Challenger* 5, 8, 10 (*Initial Reports..., 1970, 1971b, 1972*) and the later Leg 85 (*Initial Reports..., 1986a*) showed that all the north-east quarter of the Pacific Ocean eastward from the line from Hawaii to the Emperor submarine ridge originated by the out-thrusting of the ocean floor from under the North American continent during the Eocene, Oligocene and Miocene, and perhaps since the Campanian age of the Late Cretaceous (Site 63 of Leg 16—*Initial Reports..., 1973b*). The Eocene base of the sedimentary sections was traced along the 140° meridian at a distance of 1200-2500 km and more from the North-American coast; in the halfway area Oligocene base sediments of different stages were found, and near the coast, where until the Miocene, as most investigators consider, the trench existed, Miocene base sediments were observed. Although to explain this phenomenon plate-tectonically, a supposition was made about the existence of a hidden continuation of the East-Pacific Middle Ridge beneath North America (the weaknesses of such speculation will be discussed below), the fact is that the oceanic basement was out-thrust from beneath the continent, i.e., according to our terminology, it was created by eduction. This eduction process lasted at least 40, and perhaps as much as 60 million years, with a mean rate of eductional movement of the order of 6 cm per year after the middle Eocene.

The appropriate conclusion was initiated by the results already obtained in Leg 5 by *Glomar Challenger* (*Initial Reports..., 1970*): Sites 32 -36 in a deep part of the Pacific near the Californian coast established the late Oligocene—middle Miocene age of the oceanic foundation whereas 1500-1600 km to the west in the area of the 140th W meridian it proved to be early-middle Eocene (Sites 38—42 disposed along the meridian over a distance of 12 thousand km). These data were supplemented in Leg 16 (*Initial Reports..., 1973b*) where Sites 161 and 162 confirmed the early-middle-Eocene age of the ocean floor near the 140th meridian, while Site 160 which is 800 km to the east and Site 163 which is 900 km to the west established early Oligocene and early Campanian age of the floor respectively; so these explorations have yielded a latitudinal profile with eductional changes in the age of the oceanic floor relative to North America along the 11° N latitude. Several Legs that followed confirmed the age of the ocean floor near the Californian coast and California Peninsula as Miocene in most cases. The rate of the out-thrust of the oceanic floor relative to the American continent was, according to these data, at the latitude 38-40 about 6 cm per year in the Eocene-Miocene and at the latitude 11 from 3 cm per year in Late Cretaceous—early Eocene and up to 6 cm per year in Eocene—Oligocene. These data characterize the areas of abyssal plains. In the Eastern Pacific Rise, which is to the south and southeast, the changes in the age of the oceanic floor are undoubtedly connected with its different, spreading, origin and should, thus, be considered independently. Still farther to the southwest, near Central America, spreading and eduction areas possibly alternate with one another.

A supposition on a spreading ridge hidden beneath North America, considering the logical contradictions, can at first glance be confirmed by some arguments, as the northern branch of the East Pacific Rise traceable on the surface in the region of the Californian Bay is terminated by the North American coast, while the oceanic floor westward, accompanied by steady striped magnetic anomalies and large faults transverse to them, seems to be like the slope of the Rise. However, stripe anomalies in the mid-oceanic ridges are nowhere known at such large depths. The East Pacific Rise is most likely to terminate in a natural way approaching the American continent, doing so in two branches — near the Californian Bay and Central-American Isthmus. It is worthy of notice that the supposed underthrust of the spreading ridge to beneath North America took place right opposite the area of deep-sea trenches of the northwestern and western regions of the Pacific near its borders with Asia, and the width of the western wing of the supposed spreading structure here from its axis beneath North America to these trenches reaches 11-12 thousand km. The same width is to be expected for its eastern wing, now allegedly hidden; then, if it really existed, its western bound should not be closer than beneath Central Africa and, consequently, occurrence of the `ridge axis' beneath North America cannot be explained by the overthrust of the latter caused by the opening of the Atlantic Ocean. One could only assume the self underthrusting of the ridge and both of its wings eastward, from which it follows that the northwestern wing was arrested in some way at its front which hindered its underthrusting in the trenches instead of allowing its easy realization. The floor of the northeastern Pacific with a plate-tectonic-admitted eduction type of origin covers an area of more than 30 million km^2, i.e. about a quarter of the total Pacific area.

Northwestern Pacific

The scheme, to be found in a number of works, which presents a consistent increase in age of the Pacific floor from the axis of the East Pacific Rise over 12 000 km to the northwestern trenches, from the Pleistocene up to the Jurassic, is not only too rough, but proves erroneous if considered in greater detail. This scheme is valid for only the East Pacific Rise (spreading ridge) approximately within the boundaries of its manifestation in relief. At the latitude of 20° S, near the western foot of the Rise, 1000 km from its axis (the sea bottom here is of the order of 4000 m) the floor age, according to the data from Legs 92 (*Initial Reports..., 1986b*) and 8 (*Initial Reports..., 1971b*) and dating of magnetic anomalies parallel to the strike of the Rise, is estimated as early Miocene; striped magnetic anomalies parallel to the ridge are traceable for another 800 km to the west, the remotest of them, near the Tuamotu Archipelago, being dated, with the same boreholes, as middle Miocene. The Sites of Legs 7, 8 and 33 drilled from 2-3 thousand km to the northwest revealed among the most ancient deposits those of the Upper and upper part of the Lower Cretaceous (Santonian-Hauterivian) which could be considered to be a continuation of the age-gaining sedimentary cover of the East Pacific Rise towards the trenches; however, at a distance of 4000 km to the northwest from the Tuamotu Archipelago and 5000 km from the Japan and Kuril Trenches, in Site 167 of Leg 17 (*Initial Reports..., 1973c*) and disposed at the Magellan Rise in the Central Pacific Basin, the Late Jurassic (Tithonian) age of the oceanic basement had already been revealed. This confirmed the opinion suggested previously in *Initial Reports..., 1971a* (Leg 6) by B. Heezen that the Central Pacific Basin is the most ancient and enigmatic province of the Pacific Ocean. Farther, near the northeastern margin of the Central Pacific Basin in the region of the Marcus-Nekker Range (Legs 17 and 62, *Initial Reports..., 1973c, 1981b*), deposits of the Barremian-Albian and Santonian ages were again revealed. At the base of these sections, and still farther to the northwest, near the southern foot of the Shatsky Rise, in Site 307 of Leg 32 (*Initial Reports..., 1975b*), 2000 km from the Kuril Trench, the basement was again confirmed as Berriasian-Valanginian. On the Shatsky Rise, the sedimentary cover of several hundred metres in thickness, which is established on the seismographic profiles, was not completely revealed at boreholes; however, it is thought that in its base there are the Upper Jurassic deposits revealed near the western foot of the Rise at Site 49 (Leg 6, *Initial Reports..., 1971a*). Farther, in the direction of the Kuril Trench, several holes of Leg 32 which reached the ocean basement confirm the above-mentioned linear profile where the age of the floor consistently becomes younger from the Shatsky Rise towards the trench. After the Upper Jurassic deposits revealed at a distance of about 1200 km from the Kuril Trench axis (Site 49, 50, Leg 6), nearer to it at a distance of 700 km from its axis at the base of the sedimentary cover there were established deposits dated as Valanginian-Hauterivian (Site 304 of Leg 32—*Initial Reports..., 1975b*), and at a distance of 500 km as the Hauterivian-Aptian (Site 303). Unfortunately, no holes that reached the basement were drilled closer to the trench (Site 193 of Leg 19, on the Zenkevich swell, drilled through only 14 m in the Pleistocene deposits—*Initial Reports..., 1973e*).

As seen from the above data, there is nothing which would look like consistent growing of age of the oceanic basement from the axis of the East Pacific ridge to the Japan and Kuril Trenches. In the central part of the ocean there occurs an intricate alternation of the oceanic floor of the Cretaceous, including Late Cretaceous, and Late Jurassic ages. Just in front of the Kuril Trench the changes in the basement's age is of an eduction sequence in the range of Late Jurassic—Hauterivian—Aptian. As shown previously (Chapter IV) such changes within the same age range seem to have taken place to the north, in the Initial Basin of the Mesozoic Koryak margin sea which underwent folding.

Obviously, the floor of the Pacific in its different parts is of different origin and ages, which is natural for one third of the Earth's surface.

Drilling results near Mariana Trench

There is a place in the Western Pacific where the results of drilling (of the *JOIDES Resolution* vessel) were interpreted in the Report of Leg 129(*Proceedings..., 1990b*) as an ultimate success of plate tectonics' (p. v): this was because of the Upper Jurassic—Oxfordian sediments, the oldest within the Pacific Ocean, revealed near the front of the Mariana Trench at the base of the sedimentary section of the Site 801. However, in the same region, the holes in Sites 800 and 802 were also buried with oceanic basement of Berriasian and Albian—Late Aptian ages, respectively, Site 800 being closer to the Mariana Trench (500 km from the trench axis) than Site 801 (750 km; Fig. 44). So, within the area in front of this trench, it is not a subduction, but an eduction succession that is found in ages of basal sediments, if one judges

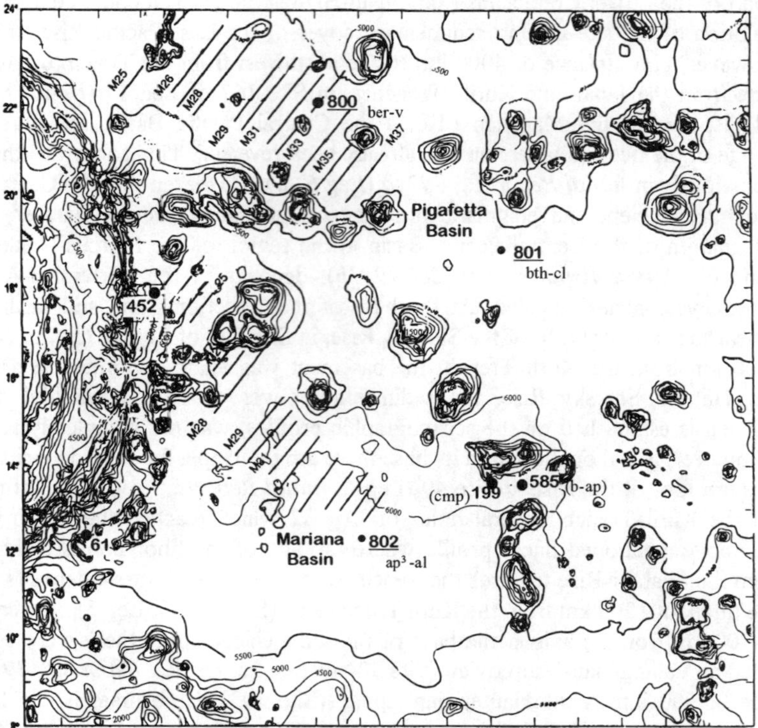

24°
22°
20°
18°
16°
14°
12°
10°
8°
146°
148°
150°
152°
154°
156°
158°
160°
162°
M25
M26
M28
M29
M31
M33
M35
M37
800
ber-v
Pigafetta
Basin
801
bth-cl
452
M28
M29
M31
585
(b-ap)
(cmp) 199
61
Mariana
Basin
802
ap³-al
5000
5500
6000
4500
2000

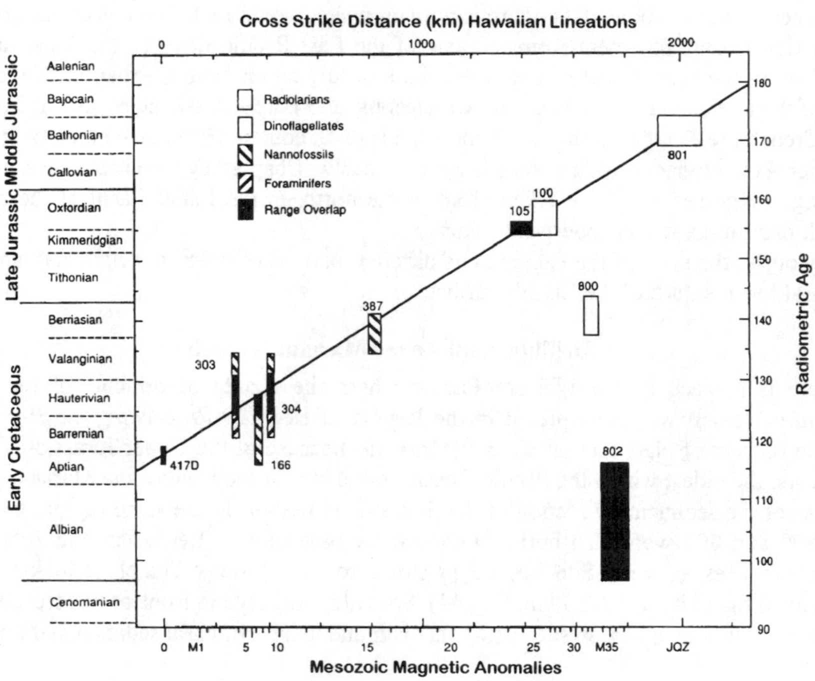

Cross Strike Distance (km) Hawaiian Lineations
0
1000
2000
Middle Jurassic
Aalenian
Bajocain
Bathonian
Callovian
Late Jurassic
Oxfordian
Kimmeridgian
Tithonian
Early Cretaceous
Berriasian
Valanginian
Hauterivian
Barremian
Aptian
Albian
Cenomanian
Radiolarians
Dinoflagellates
Nannofossils
Foraminifers
Range Overlap
801
100
105
800
387
303
304
166
417D
802
Radiometric Age
180
170
160
150
140
130
120
110
100
90
0
M1
5
10
15
20
25
30
M35
JQZ
Mesozoic Magnetic Anomalies

Fig. 44. Location of drill sites and interpretation of magnetic anomalies in front of the Mariana Trench, (*Proceedings..., 1990b,* Leg 129). **Above** location of drill sites and magnetic anomalies according to some previous decoding; ages of oceanic basement from drilling results: bth-cl—Bathonian - Callovian, ber-v—Berrisian - Valanginian, ap-al^3—Upper Aptian - Albian; in brackets, age of oldest sediments in holes which did not reach oceanic basement: b-ap—Barremian - Aptian, cmp—Campanian; at Site 452 there is Chert Clastic Suite with pure Campanian radiolarians in all cores, including those of Quaternary sediments; Site 61 is also in this suite. **Below** diagram of relation between oceanic basement ages from drilling data and from decoding of magnetic anomalies (*Proceedings..., 1990,* Leg 129); diagonal line indicates a predicted time scale constrained prior to Leg 129 on the ground of drill sites in the Atlantic and partly North-West and Central Pacific; reliability of magnetic dating for area of Leg 129 prove to be overstated.

from these two Sites. In order to explain such a disparity, the authors of the report (*idem*) have to resort to the suggestion (as was done for the Wharton Basin) of large offsets of different parts of the ocean floor, which was correlated with interpretation of poorly defined magnetic anomalies (as in the Wharton Basin). However, one can suppose that an abrupt pattern of stripe magnetic anomalies is partly typical of the ocean floor of an eduction origin; so it is doubtful that such anomalies are suitable for reliable recognition of the ocean floor age by their pattern only, without symmetrical pairs of them as within the mid-oceanic ridges. The age of such anomalies can be reliably established only by drilling; discrepancies between their preliminary dating and drilling results near the Mariana Trench are obvious and are clearly represented in Fig. 44 (from *Proceedings..., !990b).*

Site 802, disposed to the south from Sites 800 and 801, is approximately at the same distance from the Mariana trench and displays an Albian-Aptian age of the base of the sedimentary section; obviously, the ocean basement here can be included neither in an eduction nor subduction succession. It more likely belongs to the wide area on the south of the Mariana Basin (mainly, in the vicinity of the Marshall Islands) where the ocean basement is of such an age and it, probably has, a specific origin.

There is enough ground to suppose that near the Mariana Trench we see the continuation of the same band of the Late Jurassic—Cretaceous eduction which reveals itself between the Shatsky Rise and Kuril Trench and farther within the Koryak folded land. Once again, it is regrettable that the amount of drilling near the trenches is too small.

Cherty Clastic Suite of western margin of the Pacific

West of the Shatsky Rise, between it and the Izu-Bonin and Mariana trenches, there is a peculiar band hundreds of kilometres wide within which a number of Sites (Legs 20, 60, *Initial Reports..., 1973f, 1981a,* and others) revealed a specific sedimentary formation which was somewhat above the basement established in the seismoacoustic profiles; it is represented by pebbles and clastic fragments of cherty composition with clayey cement, or without it. These sediments are very difficult to drill into, and forced termination of drilling at most sites. The formation contains mixed microfossils of a wide age range, from the Upper Jurassic (in one case Kimmeridgian) to Tertiary, which have obviously been washed over. Unfortunately, the question of identification of this formation as an independent stratigraphic unit and its actual dating does not seem to have been posed; datings for individual holes vary and are often very uncertain (for example, `Albian-Cenomanian-Turonian-Coniacian-Santonian' in Site 51, Leg 6; `probably Cretaceous' but with Tertiary radiolarians in Site 52, and so on; *Initial Reports..., 1971a*). However, this is a suite, or series, of certain composition and properties, which must be comprehended as a whole and characterized by a single age dating. The latter can be rather younger than usually is supposed, as it is everywhere overlain with Late Miocene or younger deposits, dated by the microfossils they contain.

On the whole, the floor of the northwestern Pacific differs from one locality to another and seems to be heterogeneous in its genesis. Reconstruction of the development of the various sections should be carried out together with similar-age margin basins within the neighbouring Koryak folding system.

OTHER OCEAN MARGINS AND MAIN CONCLUSIONS

A lack of boreholes results in the absence of materials to draw conclusions on the movements of the ocean floor in front of the Tonga, Puerto Rico and South Sandwich Trenches. Nor do definite conclusions for Central American and Peruvian-Chilean trenches seem possible due to an inadequate number of drill holes in the vicinity of the eastern slope of the East Pacific Rise, whose materials should be treated separately. The holes drilled on the external side of the Central American Trench near its axis (Legs 66, 67—*Initial Reports..., 1982a, 1982b*) established the Miocene age of the ocean floor, and near the Peruvian Trench (Leg 34—*Initial Reports..., 1976b*) and age of Late Eocene and Late Oligocene. Determination of changes in basement age while moving away from the trench axes calls for more drill sites.

To rule out differences in opinion, it is necessary to enlarge the number of boreholes to be drilled down to the foundation at different distances from the trenches. The data available testify to the following: 1) everywhere with several representative boreholes at different distances from the trench within 1000-1500 km and over, the drilling results have established an eductional or similar succession in changes of age of the ocean floor; 2) there is no evidence available for subduction obtained in a similar way; 3) the greatest difficulties are encountered by subduction concepts with, contrary to expectations, seemingly desirable spreading ridges and trenches: each time, the arising contradictions (Pacific and Indian Oceans) call for drawing in a hypothesis of submerged ridges and their out-thrusting wings which virtually testifies to the impossibility of subduction and in favour of distribution by eduction.

Chapter VI. Geological History of the Mediterranean Region and Western Tethys in the Cenozoic and Mesozoic

The Mediterranean region comprises densely populated countries in which geological science originated and geological observations have been accumulated over a long period of time; geological structures are young here, and the geological record is, hence, very detailed—all these are factors favourable for the reconstruction of the history of this region. Numerous challenges arise, however, because the geological processes here are complicated and diverse, and are manifested in a multiple alternation of tensile and compressive processes.

Indeed, after the Middle Paleozoic, when the first oceanic Hercynian basin arose in north-central Europe and gave rise to the still developing West Tethys system, several Hercynian phases of folding were followed in the Mediterranean region by opening of oceanic geosynclinal basins (in the late Paleozoic-Triassic, Late Triassic-Jurassic, Late Jurassic-Cretaceous, and Eocene) with their subsequent transformation by folding late in the Triassic, late in the Jurassic, in the Albian, late in the Cretaceous-early in the Paleocene, and in the Oligocene. In many areas, the newly formed basins became unconformable to the previous structural framework, i.e. were superimposed. On the other hand, folding did not necessarily wholly embrace the earlier existing basins. Thus, the contemporary Black Sea is a remnant of a system of Cretaceous basins, the bulk of which underwent folding. A general regularity is a consecutive southward shift (though with a number of complications) of the newly forming basins, which trend sublatitudinally or west-north-west, parallel to the northern boundary of the present African continent.

Since late in the 19th century, nearly all tectonic hypotheses have related folding in the Alpine and Alpine-Himalayan orogenic belts to a collision of the rigid platform-type structures of Africa and Hindustan, on the one hand, and those of Eurasia, on the other. Similar views were developed by the proponents of the contraction hypothesis (E. Suess, H. Stille, L. Kober, and D. I. Mushketov), continental drift (F. Taylor and A. Wegener), subcrustal currents (O. Ampferer, A. Du Toit, and others), and deep subsurface differentiation of material (R. Van Bemmelen), and by other investigators. A. V. Peive developed the same view, based on an original conception of a differential movement of major crustal and upper mantle subhorizontal slabs. This view was also supported by most geologists, whatever their theoretical ideas, who conducted their studies in the Alps (E. Argand, E. Kraus, R. Staub, and others).

Many investigators suggested that the alternation of tension and compression in time and space is due to a reciprocal motion of Africa towards Europe and back. However, this rarely looked plausible with the Earth taken to be of a constant size.

In explaining the geological history of the Alpine region, plate tectonics was, on the whole, traditional. As mentioned above, the Alpine orogenic belt was chosen to play the role of a consuming boundary between the African and Eurasian Plates from the very conception of plate tectonic theory. The reason for this was a widespread occurrence in this belt of geologically evident manifestations of tectonic compression (Morgan, 1968; Le Pichon, 1968). All later supporters of plate tectonics put forward the same reasoning. In addition to the older problems, however, the plate concept exhibits some of its own. The first of them, already discussed in Chapter V, is common: it consists in a necessity to combine the concept of opening of the basins (accompanied by marginal trenches) with a presumed subduction of the basement of the same basins. This contradiction can be resolved only by the eduction model. The second problem is directly associated with the relationship between Africa and Europe. Its essence is that on its current northward movement Africa is abutting against the Mediterranean

Sea which has been opening up throughout the Neogene and Quaternary and contains the Hellenic Trench, a structure of small size exhibiting no indications of subduction (Chapters II and IV). Furthermore, according to one of the basic principles of plate tectonics, the plates should move across the mid-oceanic ridges along their crossing transform faults. As applied to Africa, this allows for a simultaneous motion towards west, north and east. However, the transform faults that should govern the eastward movement of Africa (distributed in the Atlantic Ocean) are greater in number; besides, this is the direction that is chosen for the African Plate by the authors of plate tectonics, since this makes it possible for the excess lithosphere of the African Plate to be consumed in the Sunda Trench (e.g. Le Pichon *et al.*, 1973), though not directly but through the Indian Ocean and Mid-Indian Ridge. Therefore, plate tectonics is once more forced to be inconsistent and combine plate motion past Europe with that towards Europe.

The suggestions that, in its most recent history, the African Plate changed the direction of its movement from eastward to northward and *vice versa* also contradict the fundamentals of plate tectonics—epecially as it is impossible to endow Africa with a property of reciprocal motion to account for an alternation of tension and compression in the Mediterranean region.

It may seem to be strange that plate tectonics, which is taken to be extremely mobilistic, experiences difficulty in explaining the collision of Africa and nearby Europe. However, this difficulty is natural in view of inevitable limitations imposed on plate motion by the assumption that the Earth remains unchanged in size.

Obviously, the only tectonic concept naturally explaining the convergence and divergence of the Gondwana and Laurasia continents is that of an expanding Earth. The controversies related to the subduction model are avoidable by using the eduction model instead. Indeed, many more possibilities arise for the continental blocks to move relative to each other over the Earth's surface undergoing substantial extension than in the case of the Earth of a constant size. Given a probable Earth extension of the order of 1.4 in the Mesozoic and Cenozoic (Carey, 1959; Hilgenberg, 1966), the vertical movement alone of these blocks could have been about 1400 km. Clearly, commensurate deviations from the strictly vertical direction could have taken place. This phenomenon now manifests itself in the different rates of opening of the simultaneously existing oceans (see also Chapter VII). A point of special importance for understanding, in particular, the geological history of the Mediterranean region is that in the case of an expanding Earth, there is no need to relate the multiple extension-compression cycles to corresponding changes in the direction of movement of Africa to and from Europe. With the expanding Earth, this alternation can be explained by periods of slowing down and cessation of a constant northward movement of the African Platform, similar to the motion of other Gondwana continents. Then the areas of earlier compression become overwhelmed by constant expansion responsible for opening of new basins, where new batches of deep-seated subcrustal material come to the surface amidst the earlier formed areas of folding structures. Therefore, the alternation of extension and compression is caused by an antagonism between local and temporary-acting compressive forces of different value and constantly acting expanding forces.

Discussed below are some features of the Cenozoic and late Mesozoic history of the Mediterranean region, with due account of the above-described views. The sequence of reconstructions is from younger to older events.

OPENING OF THE MEDITERRANEAN BASINS

Matching the Mediterranean Sea surroundings

Pushing apart of the blocks of continental crust should be regarded as the most probable cause for opening of the Mediterranean basins, with extensive appearance on the Earth's surface of a new basement of oceanic type. If this is the case, then the northern and southern Mediterranean borders can be joined together in a reconstruction similarly to the matching of the margins of the Atlantic and other oceans. Indeed, the boundaries of the northern and southern margins of the Mediterranean Sea are rather similar, though their direct matching is impaired by Corsica and Sardinia which together make up a block preventing them from convergence. Obviously, an attempt should be made at finding a different initial location of Corsica and Sardinia to allow for the convergence of the Mediterranean margins (Fig. 45).

W. Alvarez (1972) put forward an appropriate reconstruction, which is now well-known; he places the block of Corsica and Sardinia off the southern France to form a continuation of the Balear Islands, with a subsequent motion to its present position by way of its anticlockwise rotation about the northern tip of Corsica. However, this reconstruction seems to be geologically unlikely.

The Corsica-Sardinia block has a homogeneous, substantially granitoid ancient foundation and persistent internal structures whose trends are parallel to that of the islands. In the case of a north-westward rotation of the islands and their attachment to the southern coast of Europe (after Alvarez), they should have joined the heterogeneous structures of Southern Europe, whose boundaries then should have been normal to the trend of Corsica and Sardinia: these are several structural zones of the Western Alps, the Central France Massif and the Pyrenees. One cannot find signs of the extension of such structures in both Corsica and Sardinia; also few geological similarities have been found between Corsica and Sardinia, on the one hand, and the Balears, on the other. Furthermore, the Algeria-Provence Basin of the Mediterranean Sea lacks trails of movement of the Corsica-Sardinia block across its area. In the case of the displacement of Corsica and Sardinia to the north, one can also note an absence of the similarity required between the western delineations of the Apennine-Sicily Arc and the corresponding stretch of the coast of Africa between Tunisia and the Gulf of Sidra. Finally, the Corsica-Sardinia block does not completely correspond in size to this reconstruction, because the foundation of Sardinia (combined with submarine slopes, especially the extensive eastern slope) is much wider than that of the Balears. Similar arguments apply to the length of Corsica and Sardinia: together with the submarine slopes, these islands do not wholly fit into the space between the Balears and the Liguria Coast of Italy.

The author of this book put forward a different reconstruction (Chudinov, 1980), in which Corsica and Sardinia initially adjoined the Apenninic Peninsula in the east and Sicily to the south and then became detached from them in the course of opening of the Tyrrhenian Sea. Indeed, the metamorphic and granitoid foundation of Corsica and Sardinia is similar to the ancient foundation of Sicily and Calabria. Moreover, in reconstructing the fit of the northern and southern margins of the Mediterranean Sea, the structure of Corsica and Sicily is shown to be immediately adjacent to the African shore east of Tunisia. Combined with the Apenninic Peninsula and Sicily, they completely fill in the area now occupied by the Gulf of Sidra and Central Mediterranean (Messinian) Basin in the south of the Ionian Sea. Therefore, the ancient foundation of Corsica, Sardinia, Sicily and South Italy turns out to be part of the foundation of the African Platform, which is most deeply eroded in Corsica, Sardinia and Calabria.

Indeed, distinct markings of horizontal displacement of the single Corsica-Sardinia-Sicily-Apennines block relative to Africa can be inferred from a number of closely spaced north-west-trending faults found near Pantelleria Island with the help of seismic profiles across the Tunisia

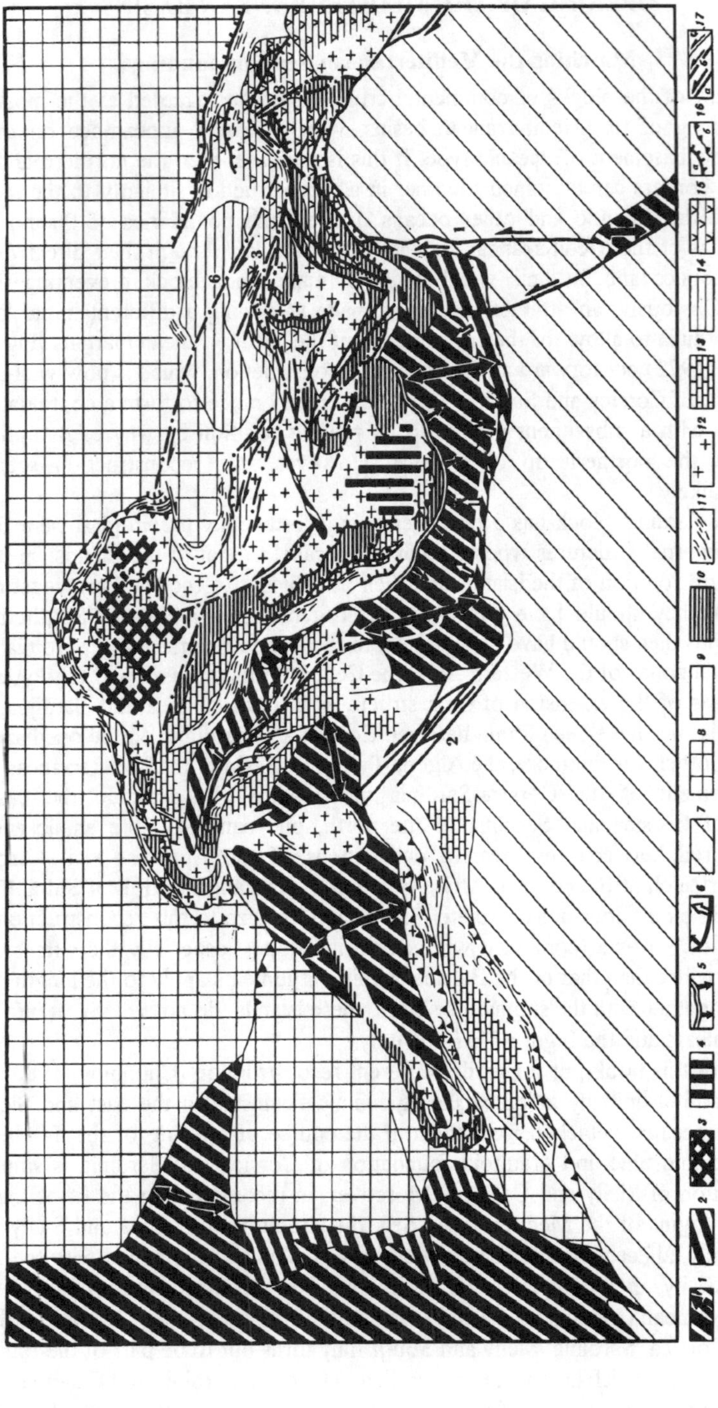

Fig.45. Main features of tectonic structure of Mediterranean region. 1—areas of new oceanic basement created in late Cenozoic through opening of the Mediterranean Sea (arrows show main directions of apart-thrust movements), also adjacent part of Atlantic floor; 2—some portions of recent continental slopes; 3—Pannon depression, area of repeated extension and block separation of heterogeneous Earth's crust; 4—Aegean depression, area of youngest extension connected with origin of the Aegean Island Arc and Hellenic Trench; 5—Padan-Levantic eduction zone of out-thrusting of oceanic basement in Miocene and earlier Pliocene; 6—Hellenic eduction zone of out-thrusting of oceanic basement in late Pliocene and Quaternary; 7—European continent between Triassic and Oligocene; 8—African continent between Triassic and Oligocene; 9—residual Black-Sea basin of late Mesozoic geosincline system; 10—South-European Ophiolite Zone with its branches; 11—mainly 'miogeosynclinal', considerably flysh, upper Mesozoic and low Cenozoic folded sediments (strokes show trend of fold structures); 12—areas of distribution of metamorphic and intrusive foundation of continental type; 13—some areas of distribution of Mesozoic continental slopes' facies (terrigenous and limestone deposits, partly gently inclined); 14—areas of distribution of mainly Paleozoic. often shallow, sediments; 15—areas of distribution of Mesozoic-Cenozoic volcanic extrusions, usually above continental metamorphic foundation; 16—main overthrusts and nappe zones: (a) connected with movement of Arabian Block, (b) others; 17—main horizontal displacements: (a) within African Platform, (b) in front of Arabian Block, (c) others. Strike-slip fault zones: **1**—Levantian, 2—Pantellerian, 3—Central-Anatolian, 4—North-Anatolian, 5—Phrygian, 6—Trabzon-Vranchean, 7—Vardar, 8—East -Anatolian.

Strait (Morelli, 1975; Fig. 46). The faults are accompanied by deformations of sediments typical of horizontal shears (e.g. Chudinov, 1962) and have a persistent trend. Evidence for probable left-lateral faulting is a northward bending of all the north-east-trending structures of Tunisia on their approach to the Tunisia Strait. According to Morelli's interpretation of the seismic profiles, the deformation in the fault system under discussion involved Miocene and more ancient sediments, whereas the slightly deformed Pliocene-Quaternary sediments overlie them with a sharp unconformity. The time of formation of the Pantellerian system of north-west-trending faults was thus close indeed to that of opening of the Tyrrhenian Sea. Because of the opening of this sea, the Sardinia-Corsica block underwent a clockwise, not anticlockwise, rotation about its northern tip. Such a rotation, greater than the reverse, is confirmed by paleomagnetic data, since most of the North Pole-pointing paleomagnetic vectors deviate southwards on Corsica and Sardinia, as compared with the Apennines vectors, but not northwards.

On the expanding Earth, growth of the opening basins, among them that of the Mediterranean Sea, is possible in different directions, including the direction of the long axis of opening; therefore, the opening of such basin as the Tyrrhenian does not necessarily lead to the shortening of the neighbouring basins, such as the Algeria-Provence. However, on the submarine slope of the Apenninic Peninsula, the opening of the Tyrrhenian Sea appears to have been accompanied by substantial compression (see below).

Following the suggested change in the location of Corsica and Sardinia, the margins of the Mediterranean fit rather tightly on the assumption that the basins of the Mediterranean and the Earth's surface occupied by their basement are non-existent (Fig. 47). In the eastern part of the sea, the formation of the Aegean Island Arc with its subsequent southward bending and of the Crete Sea behind it took place at the late stages of the East Mediterranean Basin opening, so initially the European Mediterranean border here was more rectilinear than now and thus was more similar to the form of the African border. The opening of the Mediterranean was accompanied by a wedge-like opening of the Bay of Biscay and by the well-known (and verified by paleomagnetic data) anticlockwise rotation of the Pyrenean Peninsula.

Mediterranean Basins and original sizes of the East Mediterranean Basin

The Mediterranean Sea can be subdivided (from west to east) into the following major basins: the West Mediterranean (Algeria-Provence, with its branch, the Valencia Basin), Tyrrhenian, Central Mediterranean (or Messinian, in the south Ionian Sea) and East

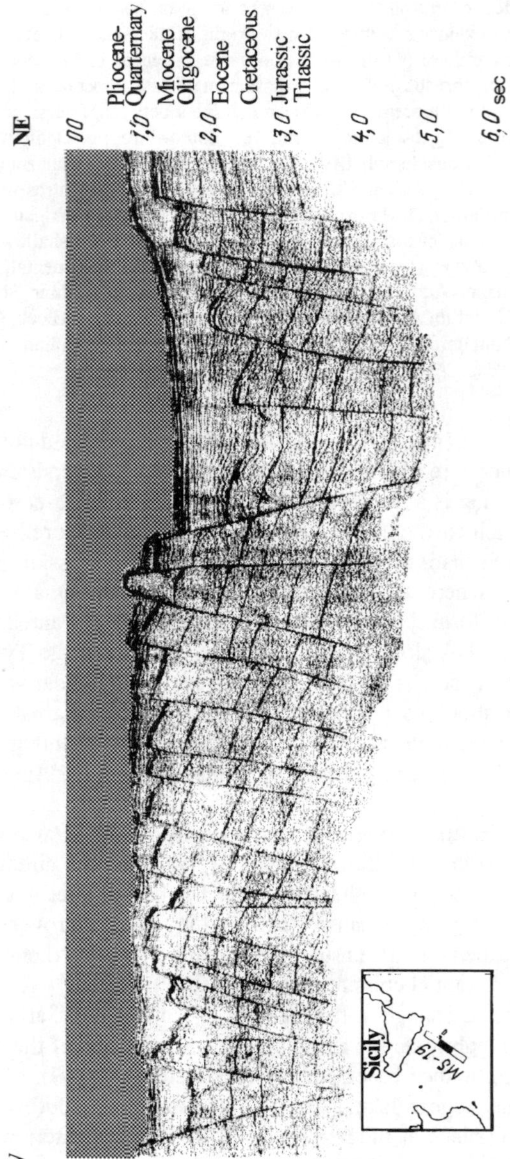

Fig. 46. Seismoacoustic profile across Tunisia Strait (Maltese graben), (after Morelli, 1975).

Fig. 46. Seismoacoustic profile across Tunisia Strait (Maltese graben), with interpretation of sediments' age, (after Morelli, 1975).

Mediterranean. Judging from the data available, the time and, possibly, mechanism of their generation are not wholly identical, but all of them arose after the Oligocene.

Reconstruction of the geological history of the Mediterranean basins requires special attention to the history of the East Mediterranean Basin. Contrary to the remaining present-day basins that have retained their size achieved, the East Mediterranean Basin seems to be larger and of some different form originally. A probable extension of the East Mediterranean Basin in the Miocene was a deep basin located in the area now occupied by the north-eastern coast of the Apenninic Peninsula, Po River Valley and the adjacent part of the Adriatic Sea (Bigi *et al.*, 1992). It is hardly possible to consider that and the present East Mediterranean Basin separately, because the latter is a clear continuation of the former along its trend, as they both existed in the Miocene. Moreover, both deep-sea areas are characterized by relics of what is apparently a single trench located in the south-western and southern margin of the initially single basin. In the East Apennines area such a relic is the Padan-Adriatic Foredeep—a deep linear structure filled in the north (in the southern part of the Po River valley) with Pliocene and Quaternary sediments as thick as 9000 m; it extends further parallel to the Adriatic Coast to the Bay of Toranto (Bigi *et al.*, 1992). In the now existing part of the East Mediterranean Basin the remnant of a trench is observed as the narrow, though not very deep (maximum depth 3489 m) Libyan Trench, which is well expressed in today's topography of the sea floor and closely follows the northern coastline of Africa. Its possible extension eastward is the Levant Trough, as deep as 3156 m. Late in the Miocene to early in the Pliocene, the East Apennine continuation of the East Mediterranean Basin underwent folding.

Eastwards, the East Mediterranean Basin originally may also have extended beyond its present boundary. This extension was destroyed by the Arabian bulk of the African Platform.

Deep-sea drilling data

Time of opening of the Mediterranean Sea Basins. Deep-sea drilling data obtained from Legs 13 and 42 A of *Glomar Challenger* and Leg 107 of *JOIDES Resolution* (*Initial Reports..., 1973a,* sites 120-134; *Initial Reports..., 1978,* sites 371-378; *Proceedings..., 1990a,* sites 650-656) are abundant enough to judge the time of opening of the Mediterranean basins. We can differentiate between the data on the abyssal parts of the basins, which can be used to estimate the time of generation of particular sections of the new ocean-type basement within basins, and the data on the slopes of basins whose sedimentary cover overlies the foundation mainly belonging to the basins' ancient margins. In the latter case, the sedimentary sequences makes it possible to estimate the whole time of existence of the basins.

Algeria-Provence Basin. A hole at Site 372 drilled on the eastern slope of the Balear Islands indicates that this basin existed in Early Miocene, as shown by the age of the most ancient sediments penetrated in the sedimentary cover overlying the Mesozoic basement of the islands. It is apparently close to the time of origin of the basin. In Messinian time, saline deposits were laid down over the entire Algeria-Provence Basin, their area having been at that time as vast as today. No attempts have been made to drill through the saline unit of the Algeria-Provence Basin (which was precluded), nor have holes been drilled in the central part of the basin. The most ancient sediments at the foot of the western slope of Sardinia (Holes 133 and 134) proved to be of Messinian age, which suggests a later involvement of this slope into sedimentation in the Algeria-Provence Basin than in the margins of the Balears.

The East-Mediterranean Basin, judging from the occurrence of lower Miocene sediments at the base of the sedimentary cover deposited on the western slope of Cyprus (Hole 376), started its opening also in Early Miocene time. Hole 129 penetrated lower Servillian sediments of middle Miocene age at the foot of the Central Mediterranean Rise. The middle Miocene sediments were also drilled by Hole 377 approximately in the same situation, in a structural cleft at the northern foot of the Central Mediterranean Rise occupying the central part of the East Mediterranean Basin. In the west of the western slope of the Central Mediterranean Rise, Hole 125 penetrated the uppermost Upper Miocene sediments containing gypsum fragments.

The Central Mediterranean (Messinian) Basin, located in the southern Ionian Sea, is among the deepest basins (as deep as 4000 m). Since it is clearly expressed geomorphologically, it should be regarded as a structure of its own value among the Mediterranean basins. The basin is situated half-way between Calabria and the Gulf of Sidra, and is characteristically triangular in shape with the semicircle northern termination corresponding to the bending of the Calabrian Peninsula. Its eastern boundary extends north-north-westwards, similarly to the structures of the East Mediterranean Basin, and its elongated southern corner is off the western termination of the Libyan Trench extending along the northern coast of Africa. The deep-sea part of the Central Mediterranean Basin—the Messinian Abyssal Plain—has been characterized by hole of Site 374 which revealed, at the base of the sedimentary section, about 40 m of Upper Miocene sediments with gypsum and anhydrite alternating with dolomite mud. Therefore, the Central Mediterranean Basin started opening in Messinian time or somewhat earlier.

The Tyrrhenian Basin is characterized by the largest number of holes (in particular, the Leg 107 drilling was conducted only in the Tyrrhenian Sea); so the time of origin of this basin is known rather definitely. The holes drilled on the wide eastern slope of Sardinia, with extensive terraces, testify to early marine sedimentation in the Late Miocene (the sediments contain gypsum and anhydrite and overlie the ancient Paleozoic basement of the island). In the deep-sea part of the basin, however, the sediments are quite thin and not older than the Early Pliocene, as evidenced by a number of holes drilled to the basement. Hence, the abyssal part (with depths up to 3800 m) of the Tyrrhenian Sea opened mainly in the Pliocene. To summarize, the opening of the Mediterranean Sea started in the Early Miocene with the opening of two main basins, the Western and Eastern, whose sublatitudinal axes, however, did not immediatelyalign: the axis of the Eastern Basin was shifted to the south. The basins of the central Mediterranean Sea—the Central Mediterranean and Tyrrhenian—originated later through the opening of the sea, in the Miocene and Pliocene. Their origin can be consistently ascribed to detachment of the Apenninic Peninsula, Corsica and Sardinia from the African Coast and their north-westward motion.

The basement of the Mediterranean basins appears to be rather heterogeneous, as follows, in particular, from the seismic data. Their analysis deserves special consideration. We only note here that several basement types can, obviously, be distinguished from these data. Direct evidence of the character of the basement of the Mediterranean Sea is still incomplete, because many holes did not reach it (mainly in areas of saline deposits).

No holes have been drilled to the basement of the East Mediterranean Basin. However, one can infer its composition from materials of its former northern continuation which is a folded East Apenninic zone. In the blocks of the north-eastern side of the Apennines, which are next to the Padan-Adriatic Foredeep, one can see a complicated mixtite (chaotic complex Argille-Scagliose and similar formations) containing abundant ophiolite fragments, and among them, some ultrabasic rocks. This suggests that the basement in the disappeared northern part of the East Mediterranean Basin was basic to ultrabasic in composition. This inference can be extended to include the now existing part of the East Mediterranean Basin. The basement of this kind can be regarded as the most characteristic of the process of eduction. Such a conclusion is consistent with the earlier mentioned probability for past existence at the south margin of the East Mediterranean Basin of a long trench, i.e. of an eduction structure. The related seismofocal zone was evidently responsible for volcanic activity in the Apenninic Peninsula and the adjacent areas.

A clearly oceanic basement was entered by a number of holes in the deep-sea part of the Tyrrhenian Sea. It consists of altered basic varieties generally cut by a carbonate veinlet system and described as metabasalt and metadiabase. According to their descriptions, these rocks resemble a basic micromelange discussed in Chapter V as a characteristic product of eduction in North-East Russia. It is quite possible that the basement beneath the Tyrrhenian Sea also contains a melange, a search of which is necessary in the present-day oceanic basement, as remarked above. In any event, among the altered basic rocks raised from Hole 651A there were serpentinized peridotites, whose occurrence was commented on in the report on Leg 107 *(Proceedings ..., 1990a_)* as `mysterious' (p. 3). But with eduction and melange, this occurrence looks natural.

An eduction origin of the basement is most probable for both the East Mediterranean and Tyrrhenian basins. As to the Tyrrhenian Sea, it is presently impossible to point to a clearly marginal structure, such as a trench, along which the eduction material could haveentered. It seems probable that such a structure could be located at the foot of the eastern submarine slope of Sardinia, but no clear seismic evidence for this is available. The seismic profiles near the foot of the Sardinia slope *(Initial Reports..., 1973a)* indicate some west-dipping faults here, but not very pronounced; there is also a deep longitudinal sediment-filled graben in the middle of the slope testifying to transverse extension.

Information about the basement in the Algeria-Provence Basin is least certain; the postulation of this basement on seismic profiles differs from that obtained in other parts of the Mediterranean. Under the sedimentary cover in most of the Algeria-Provence Basin *(Initial Reports..., 1973a, Part II)*, the basement was found to be seismically transparent, unlimited from below (to a depth of 1.5 sec and deeper), with monotonous structural complications in the form of spindle-shaped subvertical patterns, which are located at one level (0.5 to 1.2 sec below sea floor), dark-coloured on the profile records.

History of Opening

The opening of the Algeria-Provence and East Mediterranean basins, which started in the early Miocene, may be thought of as occurring in parallel. As at present, the western part of the Eastern Basin underwent a north-western bend, and the whole line of opening of both basins looked like a sigmoid curve corresponding to the modern configuration of the northern coast of Africa. The opening of the Western Basin was complicated by a partial detachment of the range of the Balear Islands from its northern margin with formation of the Valencia Basin behind them.

Initially, the fusion of the Western and Eastern basins appears to have taken place in their northern part, between the present-day Alps and Apennines, because Sicily, Corsica and Sardinia together with present Apenninic Peninsula were tightly adjacent to each other and to the coast of Africa. As the two basins were opening, the Sicily—Sardinia—Corsica—Apennines block, which originally made up the paleocontinental slope of the Eastern Basin and, possibly, the adjacent marginal continental rise, became detached from the African Platform and underwent left-lateral horizontal displacement in a north-westerly direction. As a result, the Central Mediterranean Basin—Messina Abyssal Plain—began to open up, complicating the primary structure of the Eastern Basin. Obviously, the former northern East Apennines continuation of the Eastern Basin began to be destroyed and underwent folding at the same time. Following the opening of the Central Mediterranean Basin, the now existing southern fusion of the Western and Eastern basins became possible.

After the southern restriction of the Corsica—Sardinia block had reached the northern coastline of Africa to the west, this block became detached from the Apenninic Peninsula and started to move westwards, past the African coast, with the resulting wedge-like opening of the Tyrrhenian Sea. Along with this opening, which took place in the Pliocene, the now nonexistent continuation of the East Mediterranean Basin completed its folding to become a narrow, almost suture, zone whose formation was accompanied by an extinction of the eduction zone near its former southern boundary and with the origin of a new such zone near its northern margin. This was the last episode in the transformation of the East Mediterranean Basin that gave rise to the Hellenic Trench (or, possibly more precisely, to a discontinuous system of trenches not completely formed yet), to Aegean Island Arc with its southward-convex bending, and to the opening of the back-arc Crete Basin behind the latter; behind the arc, Late Pliocene and Quaternary volcanoes arose. The opening of the Sea of Crete was probably the earliest: on the evidence provided by the Site 378, it started in Late Miocene-Pliocene time. As mentioned in Chapters II and IV, in the Hellenic Trench only undeformed Quaternary sediments have been found.

The possible eastward extension of the East Mediterranean Basin was deformed and destroyed after Late Miocene during the northward movement of Arabian Block of the African Platform (when, according to available data, the rate of its movement reached up to 10 cm/yr). The movement of this block was also the cause of the final closure of the Carpathian and Transcaucasian continuations of the Black Sea Basin, and additional longitudinal bending of the Carpathian-Rhodope Arc (e.g. Chudinov, 1980).

As mentioned above, no deformations have been found in sediments filling the very young Hellenic Trench. In the floor of the East Mediterranean Basin, however, deformations may exist associated not with subduction but with eduction. They may have manifested themselves in decrease of the width of the basin to the south of the Hellenic Trench and Crete Arc and the formation of the wide Central Mediterranean Rise, a possible wide swell in the pre-Late Miocene floor. As follows from drilling data, the Central Mediterranean Rise is, indeed, an abnormal area, to some extent. In particular, no Messinian saline deposits have been found by Holes 129 and 130 drilled in the eastern part of the rise, though they occur on the western slope of Cyprus. Also, Quaternary sediments may lie in some places on the sea basement (Site 130).

Worthy of notice are differences in the surroundings of the initial East Mediterranean and West Mediterranean basins. The surroundings of the East Mediterranean Basin were involved in more active tectonic movements, which considerably changed their structure and configuration. This is especially clearly seen in the northern surroundings, where the tectonic transformations were youngest. Volcanism was also highly developed, and in some regions (in the Apennines, on Sicily, and on islands of Greece) it was clearly of the island-arc type. The African coast east of the Gulf of Sidra is tectonically relatively calm; however, its topography was more severely dissected in the Miocene than now, as suggested by a rather complicated structural relief, with latitudinally trending scarps and grabens (e.g. Choubert and Faura-Muret, 1973) extensively buried at present beneath sands of the Libyan Desert. The Nearshore Ranges from the Cyrenaica and eastwards are its remnants. One cannot rule out the occurrence there of buried volcanic manifestations, which are eastward extensions of a discontinuous belt of young extrusions observable in Tunisia and West Libya. Opening of the West Mediterranean (Algeria-Provence) Basin may have been accompanied by a quieter situation at its margins. No marginal volcanism developed there (because the Neogene volcanism in the Central France Massif should be linked to a different structure, as explained below; volcanism on Sardinia, Sicily and in the Apenninic Peninsula is associated with the margins of the East Mediterranean

Basin, as has been discussed above). As compared with the East Mediterranean Basin, the mountainous surrounding here is more discontinuous and originated, especially in the north, in connection with older structures of the Alps, Central France Massif and Pyrenees. It seems probable that the Eastern Basin opened mainly eductionally, whereas in the Western Basin some different mechanism was at work, possibly connected with a kind of large upper mantle diapir, or some other phenomenon.

MAIN STRUCTURAL FEATURES OF THE WESTERN ALPINE-HIMALAYAN FOLDING BELT BEFORE THE OPENING OF THE MEDITERRANEAN BASINS

With the above-described reconstructed matching of the northern and southern surroundings of the Mediterranean, the pre-Miocene structure of the Alpine Belt is better understood (Fig. 47). The block of Corsica, Sardinia, Sicily and Calabria, characterized by outcrops of ancient metamorphic rocks, is adjacent to the African paleocontinent, next to its main part, the ancient African Platform. The African paleocontinent is opposed by the European paleocontinent to the north, made up of blocks with an ancient platform-type basement, among them the Russian Platform, the smaller Czech and Central France massifs, and other structures, divided and then fused by Hercynian and more ancient folding belts (these structural elements within both paleocontinents are not shown in Figs 45 and 47). The main structure separating both paleocontinents should be taken to be the South European Ophiolite Zone extending from the Betic mountain chain on the west through Liguria, the Central and Eastern Alps, Inner Dinarides and the Vardar, Subpelagonic and Pind zones in Greece and then into Turkey.

The South European Ophiolite Zone as a relic of the Central Ocean Basin of the Western Tethys

The total length of the South European Ophiolite Zone, which is a single, though at places discontinuous belt of basic and ultrabasic outcrops accompanied by deep-sea sediments, is 6000 km, and it has an extension of the same length eastwards from Turkey terminating in the Transhimalayas and Burma. The ophiolite zone proper can be taken to be mainly outcrops of the basement of an ancient ocean of the same length (since the serpentinized ultrabasic rocks are overlain in many places by sedimentary deposits of the geosynclinal type), but its width exceeds 100 km only occasionally and is frequently merely several tens of kilometres (for instance in the Krast and Miridit zones in Albania; in the Vardar zone, and in other places).
Nevertheless, many investigators rightly believe that an oceanic space whose width was many times greater (600 km, according to Trümpy, 1960); up to a few thousands of kilometres, according to Peive, 1967, 1969) used to correspond to this ophiolite belt. Narrowing took place under the action of intense folding phenomena. The maximum estimates of the original width of the ocean are more plausible, if its total length is taken into account. In this connection, S. V. Meien's (1969) data derived from the present distribution of the late Paleozoic southern and northern flora are worth mentioning: these testify to the convergence of both areas in the Himalayas to a distance of several tens of kilometres, which suggests a

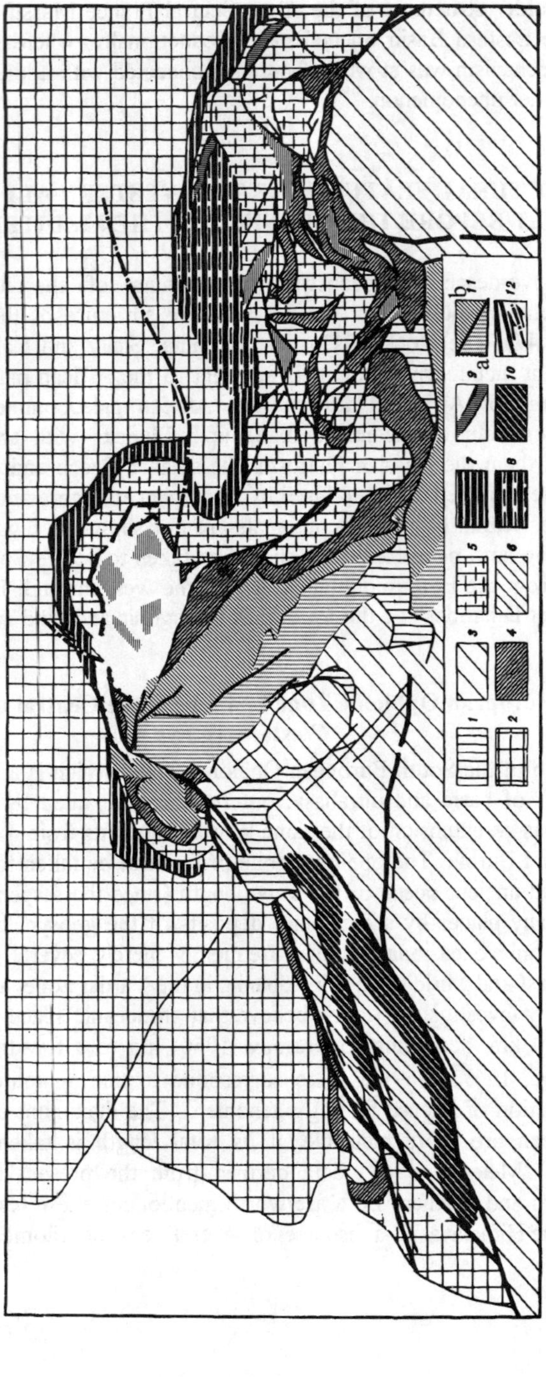
1
2
3
4
5
6
7
8
9
10
11
12
a
b

Fig. 47. Outline of matching of northern and southern surroundings of the Mediterranean Sea (pre-Miocene structures). 1—areas remaining unclosed after matching; 2—European pre-Oligocene paleocontinent; 3—African pre-Oligocene paleocontinent; 4—central area of Mesozoic-early Cenozoic Thetis Ocean, its basement and sediments having undergone universal compression in Oligocene; 5—paleo-island-arc zone derived in Mesozoic from detached parts of the European paleocontinent; 6—paleo-island-arc zone derived in Mesozoic from detached parts of the African paleocontinent; 7—northern zone of late Mesozoic-early Cenozoic marginal seas, mainly of 'miogeosynclinal' type; 8—Black-Sea basin as remnant of this zone; 9—Sevan branch of this zone with ophiolitic foundation; 10—southern zone of Mesozoic-early Cenozoic marginal seas; 11—area of Mesozoic-early Cenozoic slope of African continent in central part of pre- Mediterranean: a) upper part with wide distribution of limestones, b) lower part with predominance of terrigene sediments; 12—main faults, including great horizontal shears of Oligocene orogeny.

decrease at least of 3 to 4 thousand kilometres of the width of the Mesozoic Tethys in its east part.

As usual, geological data prove to radically contradict the fundamentals of plate tectonics, according to which oceanic lithosphere is more rigid than continental. Geological evidence has indicated for many years now that oceanic (geosynclinal) basement is, on the contrary, the least stable and most sensitive to complicated compressive deformation (see also Chapter V).

There is no question that the history of formation of the South European Ophiolite Zone was of a complex character. Being oceanic, it has been developed probably since late Paleozoic time and, no doubt, existed as such in the Triassic, Jurassic, Cretaceous and Paleogene, as suggested by fossil fauna. Sedimentation in individual parts of the oceanic basin was repeatedly interrupted by folding, which was reflected in numerous angular unconformities observed both in the ophiolite zone itself and in the adjacent areas. The proven, widespread unconformities formed in the Late Jurassic, Barremian-Albian, in the Late Cretaceous—Paleocene and in the Late Eocene—Oligocene. Some evidence shows that the tectonic deformations were accompanied by rebuilding of the general tectonic plan, more specifically by opening of new ocean floor patterns and closure of old ones. This follows, in particular, from bifurcation, in a southerly direction, of the single ophiolite band in the territory of former Yugoslavia. The resulting two branches, the Subpelagonic-Pind zone in Greece and the Vardar zone to the north from it, possibly extend farther into Turkey (as the North Anatolia ophiolite zone in the north and the Taurus zone in the south). The geological history of each zone is different. In the Vardar zone, intensely folded Jurassic and Cretaceous sediments alternate with ophiolites and are overlain by less deformed Eocene molasse-type sediments, whereas in the Subpelagonic and Pind zones, marine flysch sediments of geosynclinal type accumulated in the Late Cretaceous and Eocene. The ophiolites widespread in the North Anatolia area (among them the widely known Ankara melange) are surrounded in many places by outcropping Paleozoic and lower Mesozoic rocks and are generally regarded as older than similar rocks in the Western and Central Taurus, which are nearly everywhere associated with Mesozoic and Eocene sediments. There are reasons to believe that the northern ophiolite zone in Greece and Turkey corresponds to a more ancient (maybe originally Permian-Triassic-Jurassic) oceanic basin, whereas the Taurus-Pind area, to a younger basin, whose opening took place later and which developed most actively in the Cretaceous.

A general tendency in the opening of the Mediterranean basins was a north-to-south migration of their opening axes.

Marginal seas and island arcs at the surroundings of the Mesozoic Western Tethys Central Ocean

Probably, the zigzag plan (Fig. 45) of the opening structures in the Mediterranean region, which manifests itself in the shape of today's Mediterranean basins, arose as early as the early Mesozoic.

North of the eastern part of this zigzag (syntaxis), during the Cretaceous, Paleogene and, partially, later an elongated marine basin of the Carpathians and Black Sea existed, which approximately repeated the zigzag of the Tethys' Central Basin axis. This basin was, as a whole, parallel to the Central Basin and, being of the miogeosynclinal type, was directly adjacent to the Russian Platform. Its morphology was typical of a marginal sea, and it was separated from the Central Oceanic Basin by a chain of median massifs with an ancient foundation, among them the massifs of Slovak Ore Mountains, Bihor, South Carpathian and Rhodope in Europe; Phrygia and Galatia in Turkey, which made up together a typical island arc obviously detached from the European paleocontinent (Chudinov, 1981). In addition to the disposition of these massifs, the above said is confirmed by Mesozoic volcanism and mineralization developed within their limits. It can be concluded, therefore, that an active seismofocal zone existed beneath the above-mentioned island arc and the marginal Carpathian—Black Sea Basin in the Mesozoic and the first half of the Cenozoic; hence, the basement beneath the Central Ocean was probably formed by the eduction that accompanied that zone. The character of basement within the South European Ophiolite Band, in particular the distribution of serpentinite melange within it, is in agreement with this conclusion.

The situation seems to be different at the margins of the western part of the Mesozoic—Early Cenozoic Central Ocean now occupied by the western half of the present-day Mediterranean syntaxis. Contrary to the eastern part, the northern margin of the Central Basin is not accompanied by marginal seas and island arcs; however, possibly similar structures can be seen at the southern margin of the former ocean. In any case, this area was previously occupied by a chain of uplifts with an ancient foundation (Paleozoic in the North Morocco and Precambrian in Cabilia). This chain should be regarded as an island arc separating the Mesozoic and Eocene basins of the High Atlas and Tell Atlas (also of the miogeosynclinal type, filled, to a great extent, with flysch), and the Central Ocean. The distinction between the northern island arc area and this southern one, however, is in the lack of volcanism (which is similar to the history of surroundings of more recent West Mediterranean Basin), and also in the less diverse processes of mineralization.

Also in the Mesozoic, therefore, the oceanic basin of the Mediterranean region was divided into two, not similarly evolving, parts with the accompanying complexes of island arcs and marginal seas on the northern, European side, in one case, and on the southern, African side, in the other case. Such development of the western part of the Mesozoic Tethys once more contradicts the scenario of subduction of the single African Plate beneath the West European one. The area separating the above-mentioned parts of the Mesozoic basin is now occupied by the modern structures of the Apennines and Sardinia-Corsica.

Together with the opposing part of Dalmatia, the Apenninic Peninsula area was for long (in the Jurassic and Cretaceous) a continental shelf and slope (passive in the Mesozoic) of the African Platform. Its later partition was promoted by transformation of part of the passive African margin into an active one which accompanied appearance of the initial East Mediterranean Basin in the Miocene. This basin divided the previously existing wide African shelf into the Apennines and Dalmatia (with Apulia) parts and subjected the former to transformation characteristic of the active margin of an opening basin above a seismofocal zone.

Main phases of Mesozoic and Cenozoic folding in Western Tethys

Reverting to the overall pattern of deformations affecting the Mediterranean region, one can state that its history especially clearly exhibits extensive phases of compression that took place in the Late Cretaceous—Paleocene; Oligocene; late Miocene. These phases may be associated with an accelerated northward motion of the African Platform or its parts (Arabian Block in late Miocene time), probably with changes in the azimuth of movement. Based on the disposition of the areas and forms of the strongest fold deformations, it can be suggested that the Late Cretaceous—Paleocene (Laramian) deformations were caused by an overall north-westward or north-north-westward movement of the African Platform and the Late Eocene—Oligocene deformations, by its north-east movement.

As a result of the north-westward motion late in the Late Cretaceous and in the Paleocene, the Vardar-North Anatolia Basin was folded, longitudinal stress initiated an S-shaped bending of the Carpathian and Rhodope structures combined with partial deformation of the Carpathian Basin, and uplift episodes occurred in the Alps whose expressions are `wild flysch' and the pre-Eocene unconformity.

The change in the direction of motion of the African Platform after its cessation or slowing down in the Eocene (accompanying by opening of several new basins that inherited, to a great extent, the location of the previous basins in the Alps, Western Dinarides and the south of Turkey) led to a closure in the Oligocene and transformation into mountainous regions of the Alpine, Dinarian-Ionian, Carpathian and other basins, i.e. of the entire system of the Mesozoic-early Cenozoic Western Tethys.

The opening of the Miocene-Quaternary basins of the present-day Mediterranean Sea was complicated, beginning in the late Miocene, by a northward movement of the Arabian block separately from the rest of the African Platform. The Gulf of Aden and Red Sea opened up behind this block, and the sublongitudinal left-lateral Levant fault zone formed near its western boundary. The movement of the Arabian Block, as has been mentioned, has led to a crush of the Mediterranean structures in front of it; destruction of the Transcaucasian extension of the Black Sea; additional bending of the Carpathian Arc; initiation of a system of major, still active horizontal shears (the Anatolia Fault and others) north of the front of the block, in the territory of Turkey and the Transcaucasia; and the existence of an extensive seismic aureole now embracing the Caucasus, Turkey, the Southern Carpathians, Greece and Yugoslavia.

Volcanism and compression and extension areas

Development of the Mediterranean area was accompanied by rather extensive volcanism. Contrary to a common view of a link of volcanism to extension, the data on Southern Europe and North-East Russia suggest that the most vigorous volcanic outflows were more probably restricted to phases of compression. As shown in Chapter V, the most extensive Albian-Cenomanian effusions in the Okhotsk-Chukotka belt took place when an active seismofocal zone that existed beneath it during the Late Jurassic to the Barremian was destroyed owing to restructuring of the tectonic framework in the Aptian; so the conditions of extension earlier governing in the space above the zone gave way to conditions of compression in many places. Obviously, huge masses of molten magmas had been accumulated in the crust and upper mantle, especially under conditions of eduction, during the life-time of the seismofocal zone. beneath the Okhotsk-Chukotka belt, but did not pour out, or did so only moderately. Indeed, the rise to the Earth's surface of melts accumulated at different depths must be facilitated by application of compression additional to lithostatic pressure.

In the Mediterranean region in this connection, an extensive, though discontinuous, belt of Late Paleogene and Neogene volcanic extrusions, at places with outpourings until Quaternary time, attracts attention. This belt extends through Southern and Central Europe from the Central France Massif through Lower Rhineland, Transilvania and Rhodopes, and then into Asia Minor. Provided this belt is taken as an entity, as follows from a similar age for these extrusions, it is approximately of the same shape as the South Europe Ophiolite Zone to the south, though they are generally spaced at 300-500 km. This spacing is within the area of the usual projection on the Earth's surface of major seismofocal zones above which volcanic belts develop, for instance the Okhotsk-Chukotka volcanic belt; so a spatial association of the extrusions with the seismofocal zone, which accompanied in the north the opening of the Western Tethys in the Mesozoic and Eocene, seems to be most plausible. The age of the most volcanic manifestations known in this area, i.e. in the volcanic belt above the paleoseismofocal zone, is estimated to be in the range from the Late Triassic to the Late Cretaceous. However, most of the extrusions occurred in the Late Paleogene and Neogene, i.e. during and after the Oligocene, and partly Late Miocene, folding, when the Mesozoic and early Cenozoic seismofocal zone at the northern margin of the western Tethys became inactive.

Another example of a `delay' of volcanism relative to the time of existence of seismofocal zones can be offered by volcanism in the Apenninic Peninsula. The structure of this peninsula appears to have been located above the seismofocal zone during the existence of the primordial East Mediterranean Basin in the Lower and Middle Miocene, when it was first adjacent to the African Coast and then moved northwards, being turned into an island arc of a specific origin—because it moved tangentially, to a large extent, to the continent from which it became detached. Obviously, the genetic association of volcanism in the Apenninic Peninsula and on Sicily with a west-tilted seismofocal zone beneath them (its relic below Calabria, exhibiting deep hypocentres, is still active) must not cause doubt. But the volcanic activity in the Apenninic Peninsula was extensive in the Late Pliocene and Quaternary, i.e. only after compression and folding in the former northern Apennine continuation (from the Po River Valley to Apulia) of the East Mediterranean Basin.

As a whole, the geological history of Southern and Central Europe contains interesting material throwing light on the relationship of volcanic activity to the different stages in the development of seismofocal zones.

Can the development of Western Europe be regarded as unique?

An opinion was put forward time and again in the literature about specific, abnormal to a great extent, development of Western Europe, including the Mediterranean region. Of course, Western Europe, along with the North-East of Russia, to which it is quite similar in many aspects; the Indonesian Archipelago; and some other expanses of the Earth's surface, are some of the most geologically complicated areas. However, consideration of its history as a whole, including a comparison with that of the North-East of Russia, makes it possible to conclude that the most common regularities in the development of Western Europe are rather typical of regions where new oceanic tracts arise on the Earth.

First of all, noteworthy are constant manifestations of tensile forces that led to opening of new basins by pushing apart of the earlier existing crustal blocks and formation of a new Earth's surface inside these thrust-apart zones, with the rise of the depth material to build up this surface. This last process is effected to a large extent by the mechanism of eduction. The periodically arising areas of compression do not encompass the entire previously existing geosynclinal or ocean basins, which suggests that their origin is independent of the structure of basins and they are developed within other space patterns. It seems logical to link such forms of distribution of compressive forces with movements in the rigid surroundings of basins.

As elsewhere, folding, i.e. different deformation caused by compression, primarily involves oceanic crust making up the basement of the majority of recent and ancient seas and oceans. The South European Ophiolite Zone is one of the glaring examples on Earth of conversion of the entire ocean—more precisely part of the ocean which extended over half the circumference of the globe—into a narrow suture zone. Examples of this kind can be found in practically any folded paleo-oceanic area; therefore, the postulate of plate tectonics that it is oceanic crust and lithosphere that are the most rigid on Earth cannot be accepted.

Therefore, both in Europe and in other parts of the world, the width of postoceanic ophiolite zones in no way indicates the initial size of the basin, which can be derived only from indirect paleogeographical, paleozoological and botanical, paleomagnetic and other data on the surroundings of the basins, and partially also from palinspastic reconstructions. In contrast, the ancient surroundings of basins, especially those of platform nature, island arc structures on an ancient foundation and, to a great extent, zones of continental and island arc slopes frequently retain their initial size; so the sediments that have survived on them can be successfully used to judge the geological history of neighbouring basins, often indicating complex folding. Identification and analysis of such ancient slopes in Europe requires elaboration and have been touched upon in this book, but not in detail.

The relative proportion of areas of extension and compression in the territory of Europe (as in the North-East of Russia) does not show, at first glance, a clear predominance of the former over the latter. It should be taken into account, however, that most of the folded structures were initially extension structures similar to the existing basins. Besides, the Northern Hemisphere, especially in its northern part, is characterized by a large *piling* (A. V. Peive's term) of crustal material as compared with the Southern Hemisphere. The ratio of extension to compression areas on the globe testifies, without doubt, to a prevalence of the former both now and in the geological past.

Chapter VII. Eduction and Expansion of the Earth

Eduction, as mentioned above, when applied to the global tectonic concepts is in the best agreement with the theory of the Earth's expansion. Eduction instead of subduction on the active boundaries of the continents and oceans brings in its wake an additional growth of the oceanic lithosphere instead of disappearance of excess lithosphere formed during oceanic spreading (the formula `eduction plus spreading' as Kosygin, 1982, put it in the paper devoted to the criticism of eduction). In these conditions, one could hardly assume that the Earth's size keeps constant and reject its increase.

At the same time, eduction in itself as outflow of the depth mantle material from under the edges of continents taking no part in expansion of the Earth can be most naturally explained by expansion of the Earth's interiors producing the required internal pressure. Therewith, the distribution of eduction which should be specially emphasized, can be proved independently of the conclusions on the presence or absence of the Earth's expansion; in its turn, the evidence for the Earth's expansion can be obtained independently of assumption of eduction, in particular, only on the grounds of the data on the oceanic spreading. Thus, independently proved eduction and Earth's expansion support each other.

The hypothesis of the Earth's expansion is one of the two present branches in the development of A. Wegener's drift of the continents; the other branch is the hypothesis of plate tectonics. The Earth's expansion hypothesis has a longer history: it came into being about 65 years ago after the weakness of Wegener's hypothesis in proposing free floating of the continents over `sima'. The hypothesis of the Earth's expansion was advanced in its best known version by O. Hilgenberg (1933). He suggested that Wegener's ideas should be kept concerning the break and mutual moving away of the continents which, however, should be explained by the oceans' expansion only leading to an increase in the surface of the entire Earth. Thus, Wegener suggested that the areas of the continents and those of the oceans should be considered almost constant (and their sum strictly constant), but Hilgenberg suggested that only the area of the continents should be about constant while the oceans' area is significantly increasing. In essence, the phenomenon of spreading growth of the oceanic floor was thus predicted, but it took its actual shape on the basis of new material in the early 1960s (Dietz, 1961; Hess, 1962; Vine and Matthews, 1963) and is successfully developed nowadays. The hypothesis of plate tectonics, or `new global tectonics' which appeared later, made an effort to combine the new data on the expansion of ocean floor and the idea of the Earth's constant size—i.e. it preserved Wegener's starting idea. The problems faced by the new hypothesis along this path were not easy either.

At the same time, expansion processes observed on the Earth's surface have continually attracted the interest of investigators. They are considered mainly for oceans, but partly also for continents (for instance, Ruppel, 1995), and in some cases both oceans and continents; some authors concluded long ago that the facts prove the expansion of the Earth as a whole (e.g. Heezen, 1962, 1972; Milanovsky, 1983, 1995). Some additional evidence for Earth's expansion is presented below.

EXPANSION OF THE EARTH AS A CONSEQUENCE OF SPREADING

Le Pichon's was the only proof advanced of the impossibility of the Earth's expansion suggested by the supporters of the plate tectonics, except for general considerations about the physical impossibility of such a phenomenon, which were based on insufficient knowledge of the Earth's interior processes.

The author of this book believes that the cause of the Earth's expansion should be established after the very fact of this expansion has been proved; then the cause for expansion would have to be looked for no matter how hard this problem might be. The assertion that geology is not capable of finding out whether or not the Earth has expanded when one speaks about an increase of the Earth's surface by 1.5-2 times, is hardly acceptable.

Argument against Earth's expansion by Le Pichon

Le Pichon (1968) points out that differently oriented great circles of the globe cross a different number of the mid-oceanic spreading ridges. For example, the equatorial circle crosses the Mid-Atlantic Ridge at a rate of expansion of 4 cm/yr according to age scale of stripe magnetic anomalies; the Mid-Indian Ridge at a rate of expansion of 1.5 cm/yr and the East-Pacific Ridge at a rate of 12 cm/yr, so this makes the summary expansion of the equatorial circle of 17 cm/yr. But if one takes the meridional great circle, this will transsect only the Arctic and Circum-Antarctic middle ridges at a summary rate of expansion over different sections of from 0 to 7 cm/yr. During the last 10 my when, according to Le Pichon, no changes of the pole position took place, the equatorial circle should, in accordance with such calculation, have increased its amount by 1700 km, while the meridional circles increased by a mere 270-500 km. The Earth then would not have been round, which is justly rejected by the author of the calculation. "Consequently," Le Pichon writes, "for the hypothesis of expansion (of the oceans—Yu. Ch.) we should assume some large compensating processes of the Earth's surface contraction" (*idem*, p.107).

This reasoning is faulty because expansion of the oceans is assumed only for directions normal to the middle ridges, when actually, the oceans should be considered to be expanding in all directions, including lengthwise. This is readily convincing if one compares, for example, the length of the African coast with the length of mid-oceanic ridges around Africa, as well as the length of the Mid-Atlantic Ridge over the part which replicates the configuration of the western coast of Africa with the length of the latter. The same is true of the lengths of the continental coasts which are subject to fit with the closure of the Indian Ocean with the length of their corresponding middle ridges, and, finally, to the length of the Circum-Antarctic Ridge as compared with the length of its corresponding coastline of Antarctica. In all cases, the length of the mid-oceanic ridges turns out to be much larger (often by up to 1.5 times; in the case of the Circum-Antarctic Ridge by 1.8 times) compared with their corresponding parts of the continental coasts (as the continental slope feet). These clearly indicate the expansion of the spreading ridges along their axes too. If they are the traces of the break-up lines of the continents which have moved away, the initial lengths of the latter should have been equal to the corresponding coasts. We are going to dwell upon this again somewhat later. As for now, let us point out that the lengthwise extension of the spreading ridges reduces to nothing both the above-mentioned reasoning of Le Pichon and the pattern of displacement of plates on the whole. The spreading ridges and their accompanying stripe anomalies are a consequence of linear breaks, and the direction transverse to their strike is the most convenient for revealing the expansion traces, but it is not the only real direction of expansion.

It may be mentioned that some authors when discussing the possibility of the Earth's expansion have expressed the opinion that one can find on the Earth's surface "great circles which almost entirely traverse the continents", as, for example, V. A. Magnitsky (Magnitsky and Artyushkov, 1978, p. 107-108) have written. So one can conclude that there are circles without oceanic expansion. This is not so: one cannot find on the Earth's surface even a single great circle in which the sum of its oceanic parts would make much less than two-thirds of its full length.

Plates in new global tectonics as expansion areas

Let us now consider the consequences of spreading by accepting the postulates suggested by new global tectonics.

The plates identified in new global tectonics are known to be parts of the lithosphere for which a supposition is made about their growth in the mid-oceanic ridges on one side, and underthrusting in the subduction zones on the other. According to this scheme, plates should move from the growing borders to the consuming borders, functioning as conveyers moving towards the subduction zones. Owing to the difficulties which arise when the idea of plates is applied to the real position of the mid-oceanic ridges and trenches, an assumption is made that compensation of growth of some plates takes place not right on their own borders but in the neighbouring plates; thus, according to Le Pichon, (1968) and Le Pichon *et al.*, (1973), compensation of the African Plate expansion (i.e. of the eastern half of the Atlantic ocean) occurs not on the borders of the plate but on the consuming boundary of the Indian Plate (i.e. over the line of Sunda Trench—the Himalayas, Figs 19-20). However, the true difficulties prove to be much greater. None of six large plates identified in new global tectonics performs the above-mentioned conveyer-like movement, and each of them, if treated separately, is an expanding area.

Naturally, we do not consider numerous small `plates' which do not take part in compensation of expansion in mid-oceanic ridges and are necessary to bring plate patterns closer to the actual environment.

The conditions set by new global tectonics for the movement of the six large (main) plates are 1) spreading enlargement of the oceanic lithosphere occurs in the mid-oceanic ridges; 2) subduction can proceed in different zones identified for the purpose as trenches and fold belts. Let us add to these two self-evident conditions: 1) the overthrusting wing of the subduction zones does not disappear away from the Earth's surface; 2) if on the one side of the plate there is a border of growth and on the other side there is a border of underthrusting, then with compensation of spreading by subduction with the plate's conveyer-like movement, the distance between the above borders does not increase; if this distance increases, there is only incomplete compensation or none at all.

The African Plate is surrounded on three sides—on the west, south and north by mid-oceanic ridges, i.e. with borders of enlargement. Adjacent to it on the north is the supposed subduction zone (the Alpine belt), but it cannot make use of it, since it should move from the Mid-Atlantic Ridge from west to east along Atlantic transform faults due to which, as Le Pichon supposed, the compensation for the Atlantic expansion takes place on the consuming border of the neighbouring Indian Plate. Movement of the African Plate to the north is also hindered by the existence of the Mediterranean Sea (Chapter V). So the African Plate, if considered separately, is undoubtedly fully expanding.

The Antarctic Plate in the first scheme by Le Pichon (1968) had its own zone for subduction which was the Chilean Trench. However, in this case, the material of the plate would be forced to flow (in the horizontal direction) into the narrow `sleeve' in front of the trench and then to turn normally to its strike, which is incompatible with the postulate of the plates' rigidity. So in the later scheme (Le Pichon *et al.*, 1973), this sleeve was excluded from the composition of the Antarctic Plate, and the latter was envisaged as restricted by the single spreading Circum-Antarctic Ridge. Thus the Antarctic Plate, according to the recently accepted construction, is also a fully expanding one.

The American Plate has, for its eastern borders, the spreading Mid-Atlantic Ridge and for the western the line of trenches of the western surroundings of the Pacific. Along this line the Pacific floor can underthrust, according to plate tectonics, to beneath the American continents while the latter can only overthrust upon the Pacific. Thus, the American Plate also has no border which would allow removal from the surface of that amount of lithosphere which accumulates in the Mid-Atlantic Ridge and, consequently, this plate is also an expanding one. Sometimes an opinion is offered that compensation for expansion of the western half of the Atlantic Ocean could take place at the expense of folding in the Andean belt. To some extent, such compensation may take place, but one should take into account that Cenozoic folding deformations in most of the Cordilleras and Andes are much less than, for example, in the Alpine belt, and one cannot suppose that there is compensation here by folding for a two to three-thousand-kilometre expansion of half of the Atlantic ocean. (In this respect, one can refer to V. V. Belousov's impression of geological excursions in the Andes: "the Alpine folding in the Andes is surprisingly quiet"; the main Mesozoic trough in Chile "is structurally a wide and very gentle syncline complicated with several flexures on the wings. The Jurassic deposits overlying unconformly the metamorphic rocks of the coast zone occur almost horizontally. The above-mentioned gentle syncline is linked with the rise of the Andean Ridge through a steep flexure complicated by faults.... The Mesozoic deposits in this ridge are not at all folded"—Belousov, 1963, pp. 9-10. Such evidence is well known from other decriptions: for example, in H. Gerth's (1955) fundamental work it is emphasized that folding in the South American Cordilleras, especially in the Central Andes, is rather poorly developed).

The Indian Plate unites Indostan, Australia and their intermediate part of the Indian Ocean, limited from the southwest by the Mid-Indian Ridge. The opposite border of the plate is represented by the foot of the Himalayas on the north and by a group of trenches in front of the Solomon and New Hebrides Islands on the south, and in the middle by the Sunda Trench, i.e. by structures in which subduction is possible; however, the southeastern of them have seismofocal zones dipping southward, i.e. to the corresponding southeastern half of the Indian Plate can be applied the same reasonings as for the American Plate. As to the northwestern half of the plate, in front of the Sunda Trench there is the Wharton Basin where Upper Jurassic and Lower—Upper Cretaceous deposits have persisted (Chapter V). A question arises: if the Jurassic and Cretaceous oceanic basement is not subducted into the trenches, then how has the compensation proceeded for the expansion of the spreading Mid-Indian Ridge which is younger ? It has already been mentioned that the disposition of trenches just opposite to the spreading ridges causes, strange as it might seem, much trouble in plate tectonics construction. If one assumes that the distance from the Sunda Trench to the Central Indian Ridge has remained constant since the Mesozoic and the expansion of the ridge has been compensated by subduction in the Sunda Trench, then a lithospheric band whose width is equal to this distance, i.e. more than 3000 km, had been underthrust to beneath the Sunda Arc. This could only be in the Pre-Cretaceous and Pre-Late-Jurassic (spreading?) oceanic foundation: supposing that the rate of its formation was equal to the rate of formation of the Post-Jurassic spreading foundation, one can determine its age as the Early Jurassic—Triassic—Permian (140 + 140 my). There are no formations of this type on the opposite southwestern space from the Central Indian Ridge; nor could they be on the northeastern wing either (folded Permian—Triassic geosyncline sediments are widely distributed in the hanging wing of the Sunda seismofocal zone); but then there was nothing to underthrust beneath the Sunda Arc. If the Sunda Trench was originally joined to, or disposed nearer to, the Central Indian Ridge, then during spreading of the Indian mid-oceanic Ridge it moved back, and compensation for this spreading did not occur either. Thus, the presence of the trenches in the case of actual consideration of the spreading growth of the oceanic floor contributes nothing to substantiation of its compensation with subduction.

Thus, the trench has made no contribution to the disappearance of the oceanic basement; consequently the Indian plate also expands. If one compares the width of oceanic areas on both sides of the Central Indian Ridge, then the northeastern side, where the Sunda Trench lies will prove to be wider, i.e. it expanded, as a whole, to a greater extent. Eductionally, this is natural, since in front of this trench additional eduction growth of the oceanic crust occurred there.

The Eurasian Plate possesses extraordinary properties since it is not separated fully from the American plate; furthermore, together with the African Plate, it divides into two parts the structurally unified eastern half of the Atlantic Ocean. It has only spreading (in the Atlantic and Arctic parts) or overthrusting boundaries, so it has no room for subductioin, i.e. it is only expanding.

The Pacific Plate covers only the (larger) part of the Pacific Ocean enclosed between the spreading Eastern Pacific Ridge and trenches of its western and northwestern boundaries. This supposes a direct movement of the plate from this ridge towards the trenches (see Chapter III). The distance from the spreading ridge to trenches amounts here to 14 000 km and is wholly represented, as the plate tectonic hypothesis postulates, by spreading lithosphere which had been formed since the Late Jurassic. The whole of this lithosphere which is to the northwest of the ridge has remained on the surface, i.e. it is all area of the oceanic basement having been created by spreading. If the subduction in the trenches nevertheless took place, then in these trenches have been submerged an equal portion (i e. 14 000 km) of the pre-Jurassic ocean basement, obviously of Triassic, Permian and Carboniferous age. All geological data are in contradiction with such a construction, as they include evidence that these trenches did not exist at that time. Besides, the asymmetric position of the Eastern Pacific spreading Ridge makes the adherents of plate tectonics suppose the removal of the ridge from trenches with partial underthrusting of it under North America. As to South America, again this is so, because the East-Pacific Ridge is several times closer to it than to Asia. So, the Pacific Plate, instead of submerging in the near-Asian trenches, was held there, as if it had a support, and moved towards the opposite side. A supposition that the spreading ridge and its southwestern wing were partially superposed by both continents of America at the expense of the spreading of the Atlantic, does not change the situation since, even if both Americas moved to the east right up to the axis of the Mid-Atlantic Ridge, the position of the Eastern Pacific Ridge within the Pacific Ocean would still be asymmetric. Thus the Pacific Plate is, most likely, the most expanding 'plate' during the Mesozoic and Cenozoic.

Thus, every one of the six plates which together enveloped almost the whole of the Earth's surface, if treated in accordance with the statements accepted by plate tectonics, will turn out to be expanding. Essentially, these are conditional areas which allow the entire Earth's surface to be considered in parts (as was seen, the notion of plates itself has too little geological sense). Since all the 'plate' areas prove to be expanding, the conclusion of the expansion of the Earth as a whole is also inevitable.

Displacements of continents and mid-oceanic ridges relative to Africa

It is supposed that each pair of plates contacting over the spreading lines moves apart, rotating in opposite directions around a certain common pole (Le Pichon *et al.*, 1973, and others). This pole corresponds to the exit onto the surface of the axis passing through the Earth's center (the other rotation pole exists on the opposite side of the Earth). An idea of the axes and poles of rotation was borrowed from Euler's geometric theorem which states that the path of any point on a sphere can be described as its rotation around a certain pole; however, in plate tectonics, it is common to understand the rotation around the poles as a real path of displacements. Among numerous poles and axes determined for each pair of plates, poles and rotation axes for the African and American, African and Australian, and African and Antarctic plates have been determined.

If one follows the above-mentioned concepts and considers each pair of plates to move away from the mid-oceanic ridge which divides them, then the African continent should travel in three directions simultaneously—to the west, east and north. Since this is impossible on a constant-size sphere (in contrast to anexpanding one), the only way out is to assume that Africa keeps stationary while North and South Americas, Antarctica and Australia move away from it. But in this case, moving away from Africa should also be its surrounding mid-oceanic ridges since in the course of spreading, the distance to them should continuously increase. This is quite another movement than mutual movement of plates, and one can hardly account for independent movement of the ridges away from Africa (and each moment by precisely half the distance dividing the moving-away continents from Africa). However, the consequence of such movement of the mid-oceanic ridges is also an inevitable increase in their length since moving away simultaneously to different sides from Africa and preserving their continuity they should lengthen. We have already adduced the arguments for the real increase in the length of mid-oceanic ridges; in this case, it ensues from geometric constructions proposed by the plate hypothesis itself. An increase in the length of mid-oceanic ridges is just that phenomenon which was rejected by Le Pichon in calculating the ocean expansion rates and which led him to conclude that the Earth cannot expand.

The lengthwise extension of the mid-oceanic ridges has been found from different kinds of geological evidence. As was demonstrated by Solov'eva (1981), this growth is expressed in numerous transverse grabens which dissect the axial parts of the ridges; thus, distribution of structures of this kind are in the first place due to axial extension in the ridges, but not to their transform shears.

It is quite obvious that compensation for lengthwise extension of the mid-oceanic ridges is not foreseen in the scheme of movement of lithospheric plates and the presence of this extension distorts the basic principles of the latter which are: the equality of sizes of each plate on the spreading and subduction of their boundaries (ensuing from the postulate of their rigidity), displacement of the ocean floor only towards the transform faults, and need of compensation by subduction only for growth of oceans across the strike of mid-oceanic ridges.

The total length of the mid-oceanic ridges is almost twice as long as that of the Earth's equator, and being distributed over all surface of the Earth they form something like a framework on it. If the size of such framework increases, so does the Earth as a whole.

SOME OTHER ASPECTS OF EARTH'S EXPANSION THEORY

There are many other convincing proofs available for expansion of the Earth. For example, V. N. Puchkov (1965) cites the schemes for disposition of ancient equators on the present surface of the Earth according to Yu. M. Sheinmann, a researcher who did not share the concepts of the Earth's expansion but was most unbiased. All the Paleozoic and Mesozoic equators constructed from Sheinmann's data turned out to be displaced greatly into the northern hemisphere relative to the present equator (Fig. 48A). The significance of this phenomenon is explained by S. Carey's (1975) sketch (Fig. 48B) which presents displacement of the land segments into the northern hemisphere as a result of the Earth's expansion at the expense of the growth of the oceans situated mostly in the southern hemisphere.

One comes across assertions that the idea of the Earth's expansion is disproved by the data obtained from paleomagnetic research. But conclusions reached in recent years on the Earth's constant radius after the Paleozoic, have been based, it should be emphasized, on calculations according to M. Ward's method only (Ward, 1963, 1966); however, this contains obvious geometric errors (Chudinov, 1984). Most determinations made by various authors who used

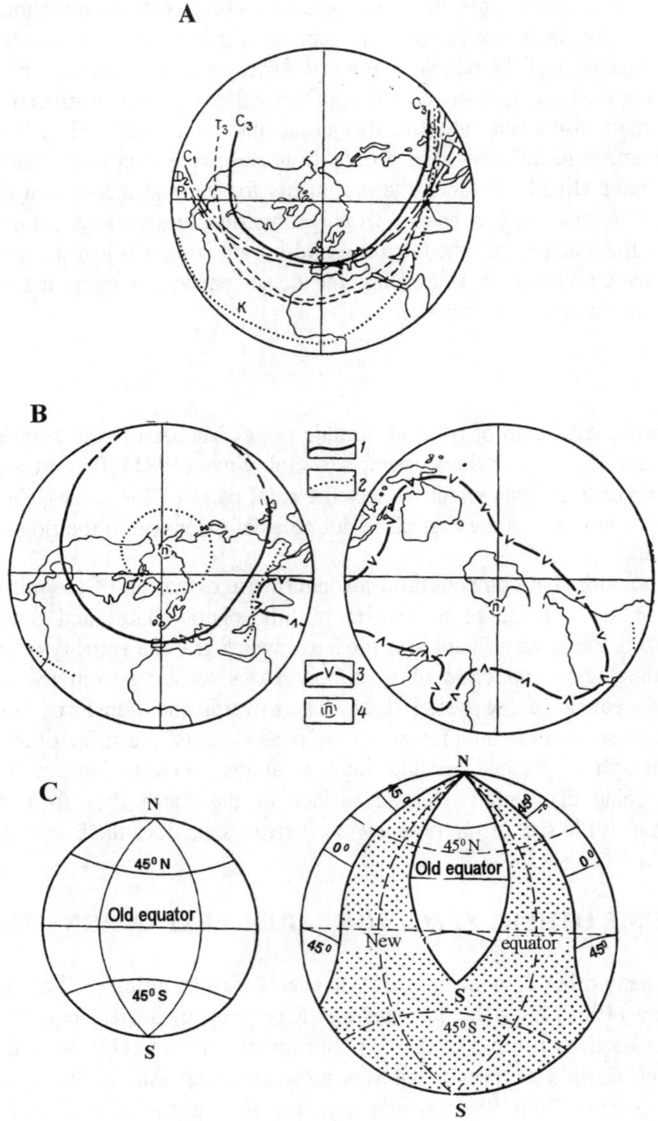

Fig. 48. A-B—disposition of equators and climatic zones in the geological past, (Puchkov, 1965) A—ancient climatic equators of Cretaceous (K), Late Triassic, Early Permian, Late Carboniferous, Early Carboniferous, Mid-Devonian (northern hemisphere); B—climatic belts in Late Carboniferous: 1—climatic equator, 2—border of arid and moderate humid zones, 3—border of southern ice zone, 4—geographic poles (northern and southern hemispheres). **C**—displacement of ancient equators northward, (Carey, 1975).

geometrically better-substantiated methods suggest a considerable increase in the Earth's sizes after the Paleozoic (with the size of the Permian radius from 0.7 or somewhat larger with respect to the present one). More exact determinations of size changes of the Earth's radius in the geological past as well as of the rate of Earth's expansion from paleomagnetic data are in principle quite possible, but are so far absent due to weak development of other ideas after emergence of plate tectonics. One of the new methods elaborated together with mathematician M. I. Tertitsky was proposed by the author of this book (Chudinov, 1984).

TWO TYPES OF OCEANS AND GEOSYNCLINES.

The hypothesis of the Earth's expansion leads us to some unfamiliar notions. Thus, when the Earth expands, a certain part of the lithosphere can move away from the other parts having extension zones both in the rear and in the front of its relative movement (although with a rate of movement which is large enough, a compression zone in the front of the movement can exist too). For the same reason, the spreading and eduction growth of the oceanic crust under conditions of the Earth's expansion, can proceed simultaneously within a single ocean. Eduction growth of the oceanic lithosphere, like the spreading growth, requires no compensation by compression (if their rates are balanced enough with the total rate of the Earth's expansion as well as with one another, and are so not much greater than the mean rate of the general expansion of the Earth).

Nature of geosynclines. There is no doubt that formation of geosynclines is easiest to explain by relying on the idea of Earth's expansion. It is now widely agreed that geosynclines (and the recent seas corresponding to them) at the initial stages of their development are extension structures. One can suppose that the Earth's expansion results in disruptions of solid lithosphere, separation of its fragments and filling of gaps between them with mantle matter rising to the surface. Because of the high density of this matter compared with the density of the continental lithosphere that undergoes disruption—to a large extent, it is continental lithosphere including that of ancient platforms—the newly formed oceanic crust has its surface sited at a rather low level (obeying the law of isostasy) and so it is normally disposed below ocean level. The geosynclinal depressions which then formed are filled with sediments whose weight make their basement descend further; an additional descent can be caused by the continuing extension of the geosyncline if it leads to greater density of its basement. Thus, formation and initial development of geosynclines are characterized by a combined manifestation of extension forces and isostasy. In `miogeosynclines', i.e. in those geosynclines whose basement is not outcropped basic and ultrabasic mantle substrate, the granite layer may be extended and thinned, which results qualitatively in the same consequences as does a full break of the continental lithosphere.

This view of the formation of geosynclines leads to rejection of the principal difference between geosynclines and oceans. Both of them emerge when the Earth's surface enlarges, their difference being mainly in sizes. If applied to present conditions, the geosynclines correspond to seas (to those of them that possess essentially basic and ultrabasic basement). From this point of view, oceans should be considered as very large geosynclines—megageosynclines. This view fits well established tectonic concepts according to which one-type oceanic basement is equally typical of marginal geosynclines and oceans, both present and ancient. The same is evidenced by a considerable arbitrariness of division of some present oceans and seas. Thus, the Philippine Sea of marginal type, which is a doubtful analogue to a number of past marginal geosynclines, is only 2.5 times smaller in area as the Arctic Ocean but if the shelves of the latter are excluded, they become equal.

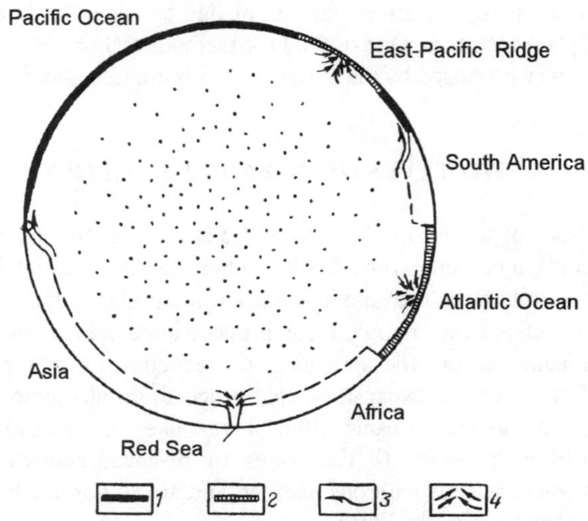

Fig. 49. Diagram of origin of different types of oceanic crust of the expanding Earth. 1—oceanic crust formed by eduction out-thrusting of mantle matter from under continents, 2—oceanic crust formed by spreading after break of continental lithosphere or previously formed eduction oceanic crust, 3—direction of mantle currents from under continents' edges to form eduction oceanic crust, 4—direction of mantle currents beneath mid-oceanic ridges in formation of spreading oceanic crust. The dotted line designates the base of the continental lithosphere, dots show internal, probably expanding, parts of the Earth.

Spreading and eduction growth of oceanic lithosphere. Both the mid-oceanic ridges and active ocean's margins with depth seismofocal zones prove to produce oceanic crust. The main difference is that in the mid-oceanic ridges, the depth matter rises directly to the ocean floor as a vertical mantle current; its extrusion onto the Earth's surface is hindered only by the oceanic crust which is thin due to its recent formation and subject to continuous rift ruptures (Fig. 49). In such conditions, a store of depth heat brought by the mantle current is delivered just to the Earth's surface and consumed in the rift zone causing volcanic and intrusive activities coinciding with the zone of highest heat flow; while both symmetrically abate on the sides. In eduction zones, the rising mantle matter has a kind of roof of colder and stronger continental lithosphere which hinders its outflow. The presence of this roof directs the mantle flow obliquely, as it has to seek an exit to the Earth's surface where the continental lithosphere is discontinued. When rising (flowing out) along the sloped lower surfaces of the continental edges, the depth matter transmits some of its heat to colder blocks. When it reaches levels where there is a drop in lithostatic pressure, the liquid and gaseous phases liberated from it seek the shortest ways to the surface and usually find them, forming through channels penetrating the continental lithosphere. During volcanism which emerges in this way, much of the depth heat is carried onto the surface directly with volcanic products within the margins of the continents or in island arcs; some of it is consumed in the areas of marginal seas (if they exist). The zone of high heat flow disposes here asymmetrically relative to the zone of appearance of the mantle material onto the Earth's surface, being displaced towards the continent. The volcanism itself is of an essentially different character compared with the volcanism of the mid-oceanic ridges; the depth matter delivered to the Earth's surface eductionally, unlike that of the mid-oceanic ridges, appears on the surface after having been cooled and is usually devoid of most of the volatile components which it had originally.

In either case, the matter being used for construction of oceanic crust does not correspond fully to the mantle's composition: under the spreading condition in the upper part of the crust there are widely distributed basalts which melted away from the ultrabasic mantle matter. Through eduction in the new-formed oceanic crust there is forthcoming mantle matter which, on the one hand, has expelled part of its components travelling across the zones of margin-continental ore-magmatic belts; and on the other, it has become enriched with the material of the continental crust due to long contact with it during out-thrusting. The different ways in which the mantle material has arrived at the Earth's surface are reflected in sharply different megastructures accompanying spreading and eduction: in the former case, this is a symmetrical edifice of the spreading ridge; in the latter, an asymmetrical more complex megastructure which emerges near the inclined surface of the mantle outflow and which has different constitution of its hanging and lying wings.

In its geophysical properties the eduction oceanic crust seems to be much like the oceanic crust of spreading origin; however, their different origins imply their dissimilarities and this issue naturally calls for special study. The eduction crust is, obviously, denser than the spreading one, since the present oceanic depths in those areas which invite the supposition of an eduction origin are greater (often deeper than 5500—6000 m) than in obviously spreading ones (usually shallower than 5500 m). One can observe higher positive Bouguet anomalies which are typical of supposedly eduction oceanic areas in the northwestern Pacific (which indicate, in the isostatic equilibrium, greater density of the oceanic basement).

The complexity of the internal structure of oceans and large geosynclines largely depend on the mutual arrangement of the eduction and spreading basement within them.

Disturbance of spreading symmetry as a consequence of eduction. It is easy to observe that, where the trenches appear, there the symmetry of disposition of spreading ridges (relative to the surrounding continents) is disturbed—the larger the distribution of trenches, the greater is this disturbance. Such a regularity can be seen in the Indian Ocean (between Madagascar and the Sunda Trench), and especially in the Pacific. The mid-oceanic ridge disposed within the Pacific—the Eastern-Pacific Rise—occupies a sharply asymmetric position, being displaced into the southeastern part of the ocean whereas its central and northwestern parts surrounded by the active type margins clearly fall beyond the mid-oceanic system.

Chapter II pointed out that such a structure of the Pacific Ocean should be considered to be a natural consequence of distribution near its western and northern borders of trenches and their adjacent areas with the oceanic floor of eduction origin. If one takes this into account, then the symmetrical structure of the spreading area in the Pacific Ocean represented by the East Pacific Rise persists. Having as its northeastern continuation the Cocos Ridge, this rise disposes not between the continents, but between areas of eduction oceanic basement, one of which is the vastest central and northwestern part of the Ocean, and another the relatively narrow southeastern one in front of the Peruvian-Chilean Trench. One can suppose that the symmetrical spreading (Paleogene-Neogene) Eastern Pacific Ridge of a mere 3-4 thousand km width dissects the province of the more ancient (Jurassic-Cretaceous) eduction oceanic basement. This feature of its formation may be responsible for the properties distinguishing it from other spreading ridges which originated at the site of the break-up of the continents: a gentler relief, a great rate of expansion, etc. In the northwestern Pacific, the eduction processes have been developing at least since the Late Jurassic; their development may have caused not only the recent island arcs but also island chains and submarine ridges in the central vicinity of the ocean.

In the Atlantic and Indian oceans, the creation of oceanic basement by eduction played a lesser part but it also manifested itself at some of their margins: in the Indian Ocean—in the area of the Sunda Trench where the deep-sea drilling data conform the eduction direction of movements of the oceanic basement, and in the Atlantic Ocean—in its northwestern near-North-American part where geophysical work has established the existence of the currently buried Mesozoic trench.

One can suppose that formation of the oceanic crust by spreading occurs more often in the oceans which is large enough (the Pacific Ocean in this case is an exception), whereas for geosynclines, especially for those which develop on the ocean margins (marginal seas), more typical is the eduction formation of the oceanic crust. In particular, as is shown in Chap. VI, eduction has widely developed on the margins of the Mesozoic Tethys. Distribution of the eduction mechanism of the creation of the oceanic basement during the latest, Mesozoic-Cenozoic, stage of the Earth's development was probably less than that of the spreading one; but it is possible that earlier the situation was different.

Boundaries between eduction and spreading oceanic crust. The question of recognizing the boundaries between the spreading and eduction crust calls for special studies. In the open ocean, when one crust is immediately adjacent to the other, this boundary apparently should be expressed by the lowering of the oceanic floor towards the area of the eduction crust, since the normal isostatically balanced position of the latter is lower. So far it is impossible to speak of any structural details of such borders; in the Pacific Ocean this may be the baseline of the East-Pacific spreading Ridge. There is also another type of boundary between the eduction and spreading oceanic crust exemplified by the meridional (of 90° E) submarine East Indian, or the Ninetyeast, Ridge. This ridge is probably a trace of the northward travel across the oceanic area of the Mesozoic Tethys of a large continental block of the ancient platform lithosphere—the Indian Platform—which was due to spreading growth of the new Central Indian Ridge (Chap. II, V). This travel became possible owing to a very high rate of the spreading opening of the Indian Ocean behind the Indian Platform: according to the paleomagnetic measurements taken in the Indian deposits, it was up to 10 cm/year (McElhinny, 1968), i.e. much higher than the possible mean rate of the Earth's expansion in the Mesozoic (up to 1 cm per year over the radius). Where the Mesozoic oceanic basement of the East Tethys was not crumbled and destroyed by the Indian block, i.e. to the east from the Ninetyeast Ridge, it was preserved in the Wharton Basin and continued to contribute to growth of the ocean foundation, by eduction, as we have seen. So the eduction oceanic crust of the Wharton Basin borders upon the spreading crust of the Central Indian Ridge, and the border is the nonvolcanic, asymmetrical Ninetyeast Ridge. Before this, the previous Mesozoic—Early Cenozoic continuation of the Wharton Basin to the northwest was obliterated in the front of movement of the Indian Platform by mighty compressive deformations and its remains were dragged away to the Transhimalayas area.

(Udintsev and Koreneva, 1980, showed that at the base of sediments revealed during deep-sea drilling on the Ninetyeast Ridge's tops, there are regularly distributed near-shore deposits of the Paleocene-Oligocene or older age which contain pollen of land vegetation brought from some location nearby. The authors believe that the sediments were deposited near Lemuria land which existed in the Indian Ocean but now is submerged; but it seems more probable that they were left after passage on the western side of the Ninetyeast Ridge of the Indian subcontinent).

Structure of the eduction oceanic crust. The mantle matter, which becomes eduction lithosphere from the time of its appearance on the Earth's surface, travels over the final part of its path, till it outcrops on the trench floor, as solid matter which had already lost its plasticity.

Over the final tens of kilometres it continuously interacts with the solid matter of the stationary continental wing, inevitably destroying it while being destroyed itself. This results in the formation, along the whole surface of the eduction out-thrust, of large masses of tectonic products which are characteristic of large fault zones, such as mylonites, phyllonites, breccias and various other tectonic mixtures, as well as of flat-oriented fault structures such as tectonic scales, lenses and wedges, together with zones of boudinage and dynamic schists. Much of this mass is transported together with the main masses of eduction lithosphere towards the opening ocean or geosyncline and is moved out onto the oceanic floor surface.

Hence there are two consequences: firstly, the eduction oceanic crust moving out onto the surface in the oceanic trenches carries in its upper part a continuous, or nearly continuous, cover of tectonically transformed matter of considerable thickness; secondly, the composition of this cover is expected to be rather variegated and heterogeneous since, besides the peridotite substrate, it can contain fragments and blocks of continental crust torn off the continental wing, including its basalt layer, as well as various tectonic mixtures. The oceanic wing matter, while being out-thrust, is subjected to regressive metamorphism since it moves towards lowering of the lithostatic pressures and temperatures; accordingly, the matter carried onto the Earth's surface displays facies of low-temperature metamorphism with signs of high pressures of a stress character. All these signs can really be detected in oceanic floor rocks at sites of probable eduction origin, as well as in the basement lying at the base of sedimentary sections in a number of ancient geosynclines where such basement is often called melange. One of the main signs of melange of this kind is its areal, not linear, distribution.

Traces of the movements experienced by the eduction crust are reflected in its inherent tectonic structures. These are first of all tectonic slabs and flakes oriented according to the direction of movement, i.e. with a gentle dip which is close to the orientation of the surface of the eduction crust, so they are often taken for subhorizontal overthrusts, etc. In general, since displacements corresponding to out-thrusting, unlike overthrusts, are not considered to be usual and therefore the possibility of their occurrence is not taken into account, the presence of the gentle-dipping tectonic contacts themselves is often taken as evidence for overthrusting. However, the age succession of deposits in the series of scales in many zones of active margins, when the scales contain sediments previous to their origin, often turns out to be reverse to the one required by the overthrusts, i.e. upper scales include younger deposits than do the lower ones. Such succession corresponds to the out-thrust direction, which is to be taken into consideration in tectonic analysis. The materials of tectonic slabs and scales in eduction crust, especially when their composition is contributed to by melange complexes, is often taken for allochtonous. This is not quite true, although they may be quite alien to the main substrate of the eduction crust: they may be detached fragments of the continental wing of the eduction zone. If eduction crust has been subjected to folding—and this is the state in which we observe it in most ancient geosynclines—decoding of the structure of such eduction crust becomes especially difficult.

Morphologically the detached fragments of the continental wing are not only flakes and slabs, but also blocks of various sizes and shapes, from fine fragments in melange to bulks making up the foundation of island arcs. Different areas of the eduction crust can be characterized by development of detached fragments of various sizes. The largest detached parts of the continents within the eduction crust are island arcs and their cognate formations. Interpretation of island arcs as oceanward moved marginal parts of continents has long been offered and maintained by many present-day researchers. This interpretation is in good agreement with the eduction direction of movements, but it contradicts that expected from subduction.

Thus, the structure of eduction crust is very complicated even before folding which distinguishes it from the oceanic crust of spreading origin, more uniform in its near-surface parts.

Differences in eduction and spreading oceanic crust in geosynclines of the geological past are important for comprehension of both their development and character of possible ore processes. The origin of oceanic crust also governs the peculiarities of later folding if it occurs.

FOLDING UNDER CONDITIONS OF EARTH'S EXPANSION

One of the common arguments against the hypothesis of the Earth's expansion is that the hypothesis is thought to be unable to account for folding, overthrusts, nappes and other manifestations of compression. However, it is hard to agree with such an opinion which is possibly based on the idea that Earth's expansion is altogether opposite to a contraction concept according to which only general contraction could give the folding effect; so the Earth's expansion is expected to result in ubiquitous tension. This reasoning ignores the fact that the only possible way for growth of the Earth's surface in the case of the Earth's expansion is the growth of the ocean surface with limited changes or preservation of areas of continents. This circumstance changes radically the consequences of the expansion of the Earth's surface.

A `great' expansion of the Earth (in distinction to conceptions of `small' expansion of several percent which does not solve the geological problems) suggests an increase of the Earth's sizes in linear dimensions by 1.4 times or perhaps some lesser but still considerable value, which is quite commensurable with the scales of spreading phenomena. Such an increase means a rise of huge near-surface masses of the Earth by one thousand or more kilometres relative to the ancient surface which might have existed at the end of the Paleozoic. This process took place in conditions of sharp inhomogeneity of the Earth's upper envelopes. This can be divided into blocks of continental lithosphere taking no part in expansion of the Earth's surface, or participating in it to only a small extent, together with oceanic lithosphere continuously increasing its area at the expense of delivery to the surface of new portions of depth mantle matter. Under such circumstances the continental lithosphere, its fragments and wreckage, inevitably hinder the outflow of depth masses to the Earth's surface, probably due to an increase in the volume of the Earth's interior. During interaction between the material rising from depth and creating the oceanic crust on the surface, and the continental blocks resisting the expansion, the former produces surface-directed one-sided mantle currents, while the latter are subjected to relative displacements by the action of strong forces. One can hardly suppose that in such conditions the rise of continental blocks to new levels was only vertical, towards the radii of the Earth, if one also takes into consideration the rotation factors, which are greater with an expanding Earth than for an Earth with constant size, and also the changes in the position of the poles on the Earth's surface which are known from paleomagnetic data.

(The question of the wandering on the Earth's surface of rotation poles is of special significance. Of particular importance are the conclusions made by Sheinmann, (1973, 1974), whose approach to the question of using paleomagnetic data was very cautious, on a high rate of pole movement which took place within some brief spans of geological time. The Earth's expansion leading to changes in mutual positions of continents and sizes of oceanic crust, may be a cause for disturbances of balance of the rotating geoid and changes in its spatial orientation. The latter in turn leads to its restructuring and the corresponding migrations of masses of the Earth's envelopes causing tectonic deformations).

So trajectories of movement of continental blocks during Earth's expansion could not be simple, as they not only moved away from one another, which was the predominant process,

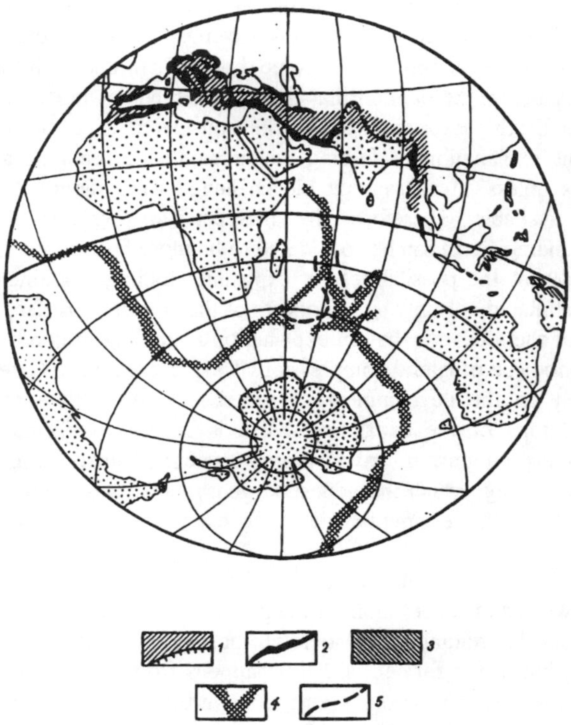

Fig. 50. Spreading oceans and Alpine-Himalayan folding belt experienced on the Earth's surface. Use is made of oblique azimuthal stereographic projection whose scale on edges is twice as large as that in centre; though these considerably reduce sizes of Indian Ocean, chart shows much smaller size of Alpine-Himalayan belt compared with oceans as well as its disposition in front of northward movement of African and Indian continents. 1—Area of deformations in Alpine-Himalayan folding belt and main overthrust structures; 2—geosynclinal zones which experienced strongest compressive deformations; 3—Circum-Pacific geosynclinal folding areas; 4—mid-oceanic spreading ridges; 5—probable initial position of southern edges of African and Indian Continents (their sizes are enlarged since after separation of continents their lines of separation in the form of mid-oceanic ridges experienced extension in different directions due to Earth's expansion); 6—dotted are continents of Gondwanaland.

but in some areas could come closer together, rotate, and occasionally collide with one another. All their deviations from radial movement led to crumpling of some parts of the weaker oceanic lithosphere which are near and between their borders. Such displacements seem to offer an adequate explanation for the observed folding phenomena and, besides, from the traditional geological standpoint of horizontal compression. In so reasoning, one should take into account that the folding areas, at any rate in any epoch of the Phanerozoic, were considerably smaller than the expansion areas, i.e. folding was less significant than the expansion of the Earth's surface. This is clearly established for the Meso-Cenozoic time from comparison of the areas of the oceans which had then been developing with areas that had experienced folding (Fig. 50).

The scale of possible horizontal displacements during the Earth's expansion depends upon the expansion scale. An increase in length of any great circle on the Earth's surface during the

Phanerozoic reaches 10-12 thousand kilometres with the rate normally taken for great expansion. Hence the subordinate tectonic displacements due to the effect of locally developing compressive forces of hundreds or some thousands of kilometres are quite possible.

From the above reasoning, the opinion that there are difficulties in explaining folding from the standpoint of the Earth's expansion should be considered as a misunderstanding. One should not forget this explanation of folding which would be recognized by a reasonable number of geologists unless an insurmountable challenge were mounted from some other tectonic hypotheses available. Therefore, the prospects for explaining folding from the standpoint of the expanding Earth can be considered to be more favourable.

It should be noted that the possibilities for explaining folding resulting from the Earth's expansion are underestimated not only by opponents of the concept but also, in many cases, by its adherents. Some of them believe that an explanation of folding in conditions of expansion must rule out the phenomena of lateral compression. Thus, according to Carey, a scientist who contributed immensely to the development of the theory of Earth expansion, folding in geosynclines in conditions of the expanding Earth is caused by formation of diapirs of undercrust matter, their bulging up over the geosynclinal surrounding and subsequent gravitational diffluence during which the deposited geosynclinal sediments move and deform; this process is considered to be simultaneous with expansion of the geosynclines (Carey, 1976).

Larin (1980) links folding with the depressive `funnels' emerging between diapirs, where undercrust matter flows in and carries along it the geosynclinal sediments which are there. One can notice, in both cases that folding is considered to be caused by plastic forcing of the matter into the geosynclines, but in the former case this happens from below and in the latter from above. A weakness of such constructions seems to be in their resemblance to many currently abandoned tectonic schemes of the 1920-1930s (for example, by Kober and van Bemmelen) These schemes imply, in particular, a concept of folding as a symmetrical process which contradicts the present-day data on the structure of the majority of fold provinces.

Real observations have long led most field geologists to the conclusion of creation of folding structures within geosynclines due to lateral pressure under active force action from the more rigid surroundings. This is, in particular, evidenced from a continuously revealed asymmetry of folded postgeosynclinal structures which develops even in the case of a symmetrical structure of a geosyncline at its initial stages.

Independence of folding processes from preceding geosyncline development

To explain folding in conditions of the expanding Earth does, however, call for rejection of some established geological views, one of them being a common view of folding as an indispensable process of completion of geosynclinal history, i.e. conditioned by the developmental mechanism of each geosyncline. The facts make us change this view since geosynclines, being areas of thrusting-apart, extension and descent, are subjected to folding only if superimposed by external compressive forces i.e. those arising outside geosynclines and caused mostly by displacement of large rigid blocks in the geosynclinal surroundings.

Thus, one can envisage that inside geosynclines, there is no kind of frequently supposed hidden mechanism providing for folding at a certain developmental stage. When outer compressive forces superimpose on the geosyncline area, the geosynclinal crust, which is rather thin and often heated and therefore only of moderate rigidity, is most readily by comparison with other parts of the lithosphere subjected to deformations (consequently the geosynclinal sediments too), and these cause development of folding just in the geosynclines. The compression fields resulting in folding are not of global but of local distribution and emerge sporadically: this is why not all geosynclines are subjected to folding and why they may undergo folding only partially. It is common knowledge that there is no span of a definite duration between the beginning of the geosyncline development and the time of its folding. There are also known instances of incomplete closure of geosynclines which were subjected to folding only locally, for example in the Mesozoic Alpine—Caucasus geosynclinal system where the Black Sea and the southern part of the Caspian Sea are regions of residual geosynclines which experienced no essential folding.

Geosynclines may not experience any folding for a long geological period and may, it seems, remain without being folded. For example, there is various evidence that, at least from the Devonian, there has existed a near-Caspian depression of geosynclinal character, the deposits in which did not undergo any folding except for that in salt diapirs. The thickness of sediments here is over 20 km, overlying, judging from its density, ultrabasic basement. It is therefore often related to the edge of the Russian Platform, which is doubtful. One should probably also regard the West Siberian lowland as an area of geosyncline nature which has existed for a long time free from folding. It can be supposed that in further geological history, any folding over many present marginal seas, whose number and area seem to grow during the Earth's development, will never occur, and the same is true of the prevailing parts of the existing oceans.

Conditions for folding

The conditions of the expanding Earth allow for two main varieties of folding: eductional and intercontinental. The essence and causes of emergence of eductional folding were expounded above (Chapter III). Some consideration was also given to development of eductional folding in the Koryak geosyncline and its surroundings which resulted in growth of new continental lithosphere on the northeastern Asian margin in the Jurassic and Paleogene.

Eductional folding develops in the hanging wing of the zones of eductional out-thrusting of depth matter due to the pressure it exerts on this wing. A necessary condition for eductional folding to develop is the presence within the hanging wing of the eductional zone of unfolded patterns of comparatively thin geosynclinal lithosphere. Most common among such patterns are marginal geosynclines which had developed earlier as a result of activities of the same eductional zone. Replacement of compression conditions by extension ones in the hanging wing of the eductional zone may be caused by changes in its inclination, in the relations between different physical phases of depth matter which can result in an increase or decrease of adhesion between material of the two wings of an eduction zone, or by other variations in general conditions. Replacement of compression conditions by extension conditions is not inevitable, and many marginal geosynclines may remain open and unfolded. If the eduction process is accompanied by folding, the eductional development leads to the creation of new continental lithosphere of folding type which builds on the active continental margins.

An example of intercontinental folding can be offered by the origin of the main parts of the Alpine-Himalayan fold belt which developed at the site of the previously existing complex geosynclinal systems of the ancient Tethys Ocean. Folding was caused by northward movement of southern continents, consisting mainly of ancient rigid platform lithosphere which, as a result, collided with Eurasia's platform-type masses. In the western Tethys, such movement was effected by Africa. in the eastern Tethys by the Indian subcontinent and by smaller fragments of Gondwanaland in the Iran-Afghanistan region. As a result, the Tethys Ocean was destroyed and turned into a folding mountainous system over a greater part of its extension.

The causes for development of intercontinental folding are sharp differences in the rigidity of lithosphere which makes up ancient continental (platform) blocks and the lithosphere of the geosynclinal provinces, and the motions of these blocks accompanied, in most cases, by spreading in the oceans behind them. Spreading appears to be a ubiquitous cause for extensive movements of continental blocks which leads to folding of an intercontinental type. The cause of migrations can be seen in a greater rate of opening of some young oceans compared with the ancient ones, with the rate of the opening exceeding the mean rate of the possible Earth's expansion.

Emergence of regions of accelerated expansion of the Earth's surface in the form of spreading-type oceans one could understand taking into consideration that their opening follows the break-up of continents, i.e. a preliminary accumulation at excessive depth pressure. It is quite possible that the rates of opening of younger oceans regularly exceeds the rates of opening of ancient oceans.

As judged by the above-obtained figures, the rates of eduction basin opening (1-2 to 5 cm per year) seem to be less than those of spreading opening. Hence deformation of the mainly of eduction origin (as shown above) Tethys Ocean due to the effect of opening of the younger spreading Atlantic and Indian oceans appears quite plausible.

The areas having undergone folding can again be disrupted and become included into new apart-thrusting geosynclines. Examples of such developments may be seen in the Mediterranean, in the area of the Persian Gulf and many others.

An indispensable condition of development of folding due to compression in the previously existing oceans and geosynclines after emergence of new (mainly spreading) oceans is the presence between the former and the latter of a separating zone containing the continental lithosphere—mainly in the form of large blocks of platform nature, etc. With a direct contact of growing eductional and spreading oceanic areas (for example in the Pacific), which move to meet each other, there does not seem to arise any considerable compression zones leading to folding. This can be accounted for by the small thickness of the oceanic lithosphere of both eductional and spreading origin, and nearness to the surface of plastic masses underlying this lithosphere whose ability to redistribute favours a rather rapid establishment of dynamic equilibrium.

In addition to the above-considered provinces there are those where manifestations of intercontinental and eductional folding prove to be combined. An example is offered by the Oloi-Anyui geosynclinal province in the North-East of Russia whose folding transformation took place within the hanging wing of the Talovka-Main eduction zone bounding the Koryak geosyncline. The basic cause of the folding was the pressure which developed in the hanging wing of this eduction zone, but the folding mechanism consisted in convergence of fragments of platform surroundings of geosynclinal structures in the previously existing Oloi-Anyui province. As seen from the above, the main sources of tectonic forces for both eductional and intercontinental folding are in the oceans or, to be more exact, in the surface-directed mantle currents generating the opening of oceans.

STRUCTURE OF CONTINENTS AND THREE TYPES OF EARTH LITHOSPHERE

The currently existing continents are probably divisible into two types. On the one hand, there are southern continents whose composition is mainly lithosphere of ancient platforms. Phanerozoic folded zones form only some subordinate fringes near their margins. On the other hand, they can be contrasted with northern continents—North American and Eurasian—whose greatest part is represented by fold provinces. Especially typical in this respect is Eurasia where the areas of distribution of ancient platform lithosphere and folding belts are about equal. The Eurasian continent is a kind of a large-scale conglomerate consisting of fragments of ancient lithosphere linked by folding provinces. One may consider Eurasia to be a new type of continent of special inhomogeneity, whose formation became possible only in the Phanerozoic under accelerated expansion of the Earth. The present continents continue to experience separation due to extension—the southern continents in the zone of the Great African faults, and in the Amazon River basin; and Eurasia, in the regions of the Mediterranean, Lake Baikal, and elsewhere.

Throughout the Phanerozoic, especially during the latter half of it, there was a continuous rise in the level of the continent relative to the oceanic surface. This rise is clearly seen from distribution over all continents of pre-Cenozoic sea deposits of different ages, not only in the fold provinces but also in their subhorizontal occurrence over vast tracts of ancient platforms. The traces of a general rise in the continent's standing are widely eroded subhorizontal beds of platforms and highly elevated surfaces of peneplains. The cause for a relative rise of continents seems to be a general descent of the World Ocean level connected with the Earth's expansion. Thus, the area of dry land on the Earth increased during its expansion.

Along with ancient platforms and folding provinces, some continents seem to comprise unfolded geosynclinal structures if they, due to relatively slow opening and abundant clastic product, are filled with sediments with an accumulative surface located above sea level. A rise of such accumulative surfaces from beneath the sea level is also favoured by descent of the World Ocean surface relative to the continents as geological time passes. In the author's opinion, as has already been mentioned, the Near-Caspian depression and West Siberian lowland, and also Pechora, Khatanga and some other depressions of the Eurasian continent represent such geosynclines. One can probably also regard the areas of the Great Plains in North America and the Upper Amazon depression in South America in the same way. Horizontal occurrence or a gentle dip of sediments in such areas is often taken as evidence of their platform nature and the existence of a rigid base beneath them. From the view presented here it is only a consequence of their development outside local fields of intense compression whose superposition could have led to distribution of folds over all these areas. Just the same, the sediments in expanses of oceans and young seas are also unfolded despite the presence of oceanic lithosphere beneath them that was undoubtedly capable of being folded. It should be supposed that the gentle dip of the sediments can be considered as a sign indicating the existence beneath them of a rigid basement only if it is attended with signs of a rather high hypsometric standing which is characteristic of the ancient platforms, also of a limited thickness of the sedimentary cover and discontinuities in it. Submersion of the basement to a depth of 10-12 thousand metres as is the case in the northern West Siberian lowland and above 20 000 m in the Near-Caspian depression is most likely evidence oftheir geosynclinal nature.

The above discussion suggests that three types of lithosphere may be distinguished: 1) ancient continental ***platform-type lithosphere***, 2) ***oceanic (geosynclinal) lithosphere***, and 3) ***oceanic folded lithosphere*** transformed by folding.

Ancient continental (platform-type) lithosphere

The emergence of such lithosphere from the standpoint of the expanding Earth can be imagined unambiguously enough. Rapid expansion typical of the Mesozoic and Cenozoic could not proceed throughout all geological history. Obviously, the continental platform lithosphere, consolidated down to a depth of several hundred kilometres, formed during a span of geological history which preceded the Phanerozoic and Upper Proterozoic and lasted several billion years in conditions of rather slow or no expansion of the Earth. Formation of this lithosphere depended on slow-occurring differentiation of the Earth's matter of gravitational and other types, depth crystallization and other processes. This created a rigid envelope over all the Earth. An increase in the rate of the Earth's expansion resulted in a progressive break of this ancient envelope. Currently it is represented by fragments of various sizes: from the largest

African Platform extending from north to south over 7000 km, to variegated `median' massifs within continents of hundreds of kilometres in size. The signs of ancient continental lithosphere are as follows: 1) a high hypsometric level of standing of its surface exceeding in most cases the present level of the ocean, 2) small thicknesses and discontinuous structure of the sedimentary cover, 3) weak deformations of sediments of "epeirogenic" or of moderate and local block—fold-block types.

Oceanic (geosynclinal) lithosphere

The wide distribution of oceanic lithosphere which is increasing with geological history is due to the late stage in the Earth's development characterized by acceleration of general expansion of the Earth and dividing of ancient continental lithosphere into their large fragments—platforms. In oceans (megageosynclines) and geosynclines which arose among the fragments of the ancient lithosphere there formed new thin (tens of kilometres) lithosphere of basic-ultrabasic composition. As was shown, there are two mechanisms for formation of such lithosphere—eductional and spreading. The latter is currently most widely distributed, but in the past, their relations might have been different. Oceanic lithosphere of both eductional and spreading origin is not very strong and easily deformable in collisions, rotation and other displacements of blocks of ancient continental lithosphere followed by creation of local compression fields. Many oceanic and geosynclinal areas do not experience any folding, and distribution of such areas seems to be increasing with geological history. A constant sign of lithosphere of geosyncline type which is due to isostasy is a low hypsometric standing of its surface: 5.5-6.5 km below ocean level for eduction-type lithosphere and mostly less than 6-6.5 km for spreading lithosphere. When the geosyncline is being filled with sediments the surface of its basement can subside to a depth of 20 km or more.

Folded oceanic lithosphere

During superposition of the compression fields and their transformation due to folding processes, oceanic and geosynclinal lithosphere develop greater thickness with corresponding decrease of area, and internal structures get more complex. Under these phenomena and, occasionally, due to developing magmatism and metamorphism, the oceanic lithosphere consolidates, which means that it becomes stronger and more rigid compared with its initial state. The standing level of the folded lithosphere surface is normally much higher than that of the level of the ancient platforms. In most cases the folded lithosphere, which emerged from the oceanic one, turns into a mountainous province which should be taken as an isostatic consequence of its thickening.

The folded lithosphere which emerges in this way is essentially different in nature from the lithosphere of ancient platforms. They differ in the lifetime and essence of their developing processes and, consequently, in structure, composition, thickness, mechanical properties, etc. Ancient platforms are a product of slow evolutional growth of the lithosphere in comparatively unified conditions of the early stages in the Earth's development, while the lifetime of the Late Pre-Cambrian, Paleozoic and the younger folded lithosphere is much shorter; besides, the younger folded lithosphere resulted from destruction and transformation of the oceanic lithosphere whose origin is quite different from that of the platform lithosphere, and was formed in conditions of accelerated expansion. Thus it should be concluded that there are no possibilities for folding lithosphere to transform ever into lithosphere of platform type, as it is often supposed.

Therefore, from the standpoint proposed, a gradual evolutionary transformation of folded lithosphere into continental platforms is impossible. Depth reworking of the roots of the folded lithosphere during geological time does take place, but it is apparently a quite different process.

As to the combining of the folded and ancient platform types of lithosphere within the continents, it does not proceed on the basis of their genetic similarity and gradual transformation of the former into the latter, but is due to the high (i.e. in relation to the dry land) hypsometric position of both of them and to spatial connection of folding with the borders of the platform massifs from whose side the pressure to generate folding is developed. (One of the confirmations for genetic heterogeneity of platforms and folded provinces is a sharp nonconformity of the borders of folded provinces adjacent to ancient platforms with the latter's internal structures; this has long been pointed out by Yu. M. Sheinemann. Reiteration of strikes which substantiates the idea of gradual overgrowing of ancient platforms with the Post-Early -Proterozoic folding systems can be observed only on the borders of different-age folding systems).

This conclusion about the three types of Earth lithosphere agrees with available geological and geophysical data. The most usual geological divisions of the present Earth's surface into ancient shields (also cratons), mountainous (orogenic, folding) provinces and oceans also correspond to these types.

The issues of the origin of the Earth's crust, of the nature of the Mohorovicic discontinuity, of the essence of geophysical boundaries in the upper mantle and other general geophysical and geological problems should obviously be solved for each type of Earth lithosphere individually.

Chapter VIII. Eduction and Mineralized Belts.

The real existence of the eduction process is confirmed by actual material obtainable in ancient geosyncline areas and also from deep-sea drilling. Eduction outflow of mantle matter is a natural consequence to the Earth's expansion and it can thus be coordinated with the general theory of the Earth's expansion: this relies on independent evidence, in particular, on materials of oceanic spreading and paleomagnetic data.

Of special practical importance is the explanation for formation of mineralized belts developing above seismofocal zones. Spatial connection between mineralized belts and seismofocal zones, both present-day and ancient (paleoseismofocal), can be considered an established fact. The young, mainly Cenozoic, belts with pronounced concentration of endogenic mineralization are disposed above the recent seismofocal zones near the Pacific coast of South America, in Central America, Japan, the Philippines and other localities of the Pacific surroundings. Beneath the wide mineralized belt of western North America there is now no acting seismofocal zone but it existed rather recently, which is testified to by the buried trenches near the western coast of this continent (Scholl, 1974).

The mineralized belts of the Pacific surroundings have long been identified by many researchers as a special Pacific type extending very far into the continents around the Pacific. (The Okhotsk-Chukotka mineralized belt with the paleoseismofocal zone beneath it in the Mesozoic geological past, existed beneath this belt, the Mongol-Okhotsk belt, according to S. S. Smirnov, and others). However, there are grounds to suppose that the belts of endogenic mineralization of the same nature were also common outside the Pacific regions. For example, the Appalachian, mainly Mesozoic, mineral belt formed above the Mesozoic paleoseismofocal zone which outcropped onto the surface on the western margin of the Atlantic Ocean, which is also evidenced from the trenches buried by Post-Cretaceous deposits and found here by seismological methods (Emery, 1965). As shown in Chapter VI of this book (see also Chudinov, 1980), the history of the Mesozoic Tethys can be interpreted as an opening of the oceanic megageosyncline attended on both sides, especially in the north, by a chain of island arcs and marginal seas with their corresponding paleoseismofocal zones. It is above these zones that most mineral regions of the Mesozoic-Cenozoic age formed in South America, Turkey and Algeria. Of a similar origin, no doubt, are many mineralized regions in the northern surrounding of the Tethys in Asia (in Indochina, China, etc.). Over the territory of the former USSR, paleoseismofocal eduction zones with their corresponding mineralized belts obviously accompanied the northern Paleozoic Chu-Ily and South Tien Shan geosynclines. Literature sources cite the well-founded suppositions of the presence of the Middle—Late Paleozoic paleoseismofocal zone beneath Kalba and Rudny Altai. In all likelihood, there existed comparable seismofocal zones which caused development of mineralization in Australia, South Africa, West Europe, etc.

It should be concluded that the Pacific type of mineralized belts can only be identified as a specific variety of the world above-seismofocal mineralized belts which concentrate the world's majority of resources of tin, molybdenum, tungsten, antimony, mercury, bismuth, most resources of copper, lead, zinc and many other metals.

Mineralization from the eductional viewpoint. It is natural that the supporters of plate tectonics paid attention to the connection between the position of mineralized belts and deep seismofocal zones, i.e. subduction zones, in their treatment. Works devoted to this issue including discussion of its practical and theoretical aspects made their appearance soon after the publication of the fundamental papers dealing with plate tectonics, but the ore-formation line failed to develop into a substantial branch of plate tectonics and enjoyed few publications, compared with other trends. Besides, no significant development of the initial conclusions (e.g. Sawkins, 1972; Sillitoe, 1972, 1976) has been observed.

Many difficulties arise in the subductional interpretation of the formation of above-seismofocal zones. First of all, there are those involved in explaining the sources of mineral matter. Most common is an opinion that the ore material is brought into the subduction zone from the underthrust oceanic plate. However, the oceanic floor (sediments and basement) is not ore-bearing everywhere, but the mineralized belts seem to emerge above all seismofocal zones. Furthermore, whereas the oceanic deposits are known to contain, even in commercial quantities, lead, zinc, copper and some other ore compounds, these deposits were not found to contain rich concentrations of such metals as tin, tungsten, antimony, and mercury, which are typical of above-seismofocal mineralized belts. Nor is it likely that an explanation would be found for the mechanism of underthrusting of ore-bearing surface sediments down to a depth of 600-700 km. Finally, each and every hypothesis dealing with mineralized belts should explain the cause of ore zonation which is observed transverse to the strike of the belts, or it should substantiate the principal possibility for this. This zonation, i.e. distribution of deposits of various metals over different distances from the trenches, is not totally similar within different mineralized belts, but it is invariably present. It is hardly possible to suppose that the underthrust ore-bearing sediments are distributed throughout all subduction zones not only evenly but also regularly so that they should ensure mineral zonation, even if it existed in the oceanic sediments before they underthrust.

There is another opinion— that of mobilization of mineral matter in the hanging wing of the subduction zones under the effect of subduction-accompanying magmatic and metamorphic processes. In this case, ore matter is not brought by subduction, and zonation seems to be easier to explain. However, this poses a question of the energy sources ensuring metamorphic reworking of the whole volume of the lithosphere above the above-seismofocal zone and removal of metals from their initial locations with their vertical transfer towards the Earth's surface over distances of hundreds of kilometres. As was already pointed out, the subduction concepts are most challenged in explaining the origin of heat which results in distribution of magmatic processes and high heat flows above the seismofocal zones. The amount of heat required in the corresponding calculations (which are rather debatable) are taken as the minimal necessary; reworking of the above-seismofocal lithosphere to redistribute mineral components on such a large scale requires much additional energy.

EDUCTION MODEL OF ORE FORMATION

Along the surface of eductional out-thrust geophysical mantle matter rises from depths of 600-750 km. According to available calculations, mantle at this depth is as hot as 2000-2300° C , or more, and is at a lithostatic pressure of more than 200 kilobars. Rising along the inclined eduction surface this matter goes over levels of successively lowering lithostatic pressure while retaining its anomalously high temperature; a natural consequence of such outflow is a rise in temperature of the lithosphere above the eduction zone along the entire path of educting out-thrust material (except for its uppermost part) and separation of the most mobile components from the rising masses, by means of their partial boiling in particular.

To specify how the process develops is so far rather difficult, mainly because of scarce data on the composition and initial physical condition of the mantle at depths of about 700 km and others. It is known, however, that the level of 600-700 km, i.e. the depth where eduction zones emerge, correlates with that of the principal refraction in the curve of seismic wave velocities;

deeper that this, the increase in velocities of longitudinal and transversal waves with depth drops about four-fold. The span above this level, between it (or 650-1000 km depth) and 400 km (Golitsin's discontinuity), is often considered to be a transitory layer from the upper to the lower mantle. Since the existence of seismofocal zones encompassing this layer and terminating at the Earth's surface, unites it with the upper part of the mantle, there are grounds to consider the level of about 700 km where seismofocal zones are engendered to be the base border between the upper and lower mantle. At any rate, the lower mantle (Bullen's 'D" layer), which continues down to the core border, according to the data from an increase in velocities of seismic waves begins from the level of about 700-1000 km. In accordance with the above-expounded pattern of the structure of the Earth's lithosphere (Chapter VII), it can be supposed (and this does not seem to contradict geophysical data) this border is inherent only in ancient continental (platform) lithosphere and is absent (at such levels) beneath the oceans and fold areas. In this case, it is the matter of the lower mantle originally disposed beneath the ancient platform's within the continents which undergoes eduction and then appears on the Earth's surface.

One can suppose that the original composition of this material was ultrabasite but differed from ultrabasic rocks found on the surface in their content of components lost during the eduction rise. From this viewpoint, the `alpinetypic' ultrabasites which largely are the depth matter out-thrust eductionally are an eduction remnant. Among other lost components which were present in the composition of the initial lower mantle matter in greater quantities than they are found in ultrabasites on the Earth's surface, there could also be ore-forming compounds. This supposition looks plausible since the material which is at depths of 600-7000 km and deeper should be weakly differentiated. In particular, it could contain rather heavy concentrations of different metals, sulphur, and other elements which, during eduction rise of the lower mantle matter to the surface, separate from it as parts of mobile compounds and infuse into the hanging wing of the eduction zone. This is evidenced by the composition of chondrites, which are usually correlated with the lower mantle matter and, indeed, contain considerably heightened amounts of many elements typical of the above-seismofocal ore belts.

In eduction, the ore compounds are thus supplied by mildly differentiated matter of the lower mantle during its outflowing and unloading in conditions of successively decreasing pressures while preserving the high temperatures characteristic of depth. Such explanation meets all the demands which could be extended to a tectonic model describing the origin of ore belts. So the eduction model allows an explanation of all main features of above-seismofocal ore belts: their disposition above seismofocal zones; alternation of different ore zones transverse to the strike of the belts (as a consequence of changes of unloading conditions of mantle matter during its way over different depth levels); independence, or a limited independence, of the character of ore formation and ore composition from the composition and structure of the Earth's crust within ore belts (Shilo *et al.*, 1979). The last circumstance seems to testify to the depth (mantle) origin of the ore matter in mineralized belts related to the active margins of the continents.

The most valuable ore deposits of the above-seismofocal mineralized belts are large massifs of rocks which underwent sulphide or other ore mineralization over the entire volume. In this case, the main ore bodies are not veins or replacement zones along the fractures but ore-impregnated masses or rocks overfilled with irregular mineralized veinlets of stockwork type. It has always been a challenging task for the application of hydrothermal theory to explain the formation of such deposits, especially of their varieties, which are characterized by rather uniform impregnation of large volumes of rocks with ore compounds.

A supposition on the deposition of ore matter from gases came to life before the hydrothermal hypothesis and has been developing, along with the latter, for more than 200 years now. A very interesting review by S. S. Smirnov (Smirnov, 1945), well-known in Russia, considered these two hypotheses as equally plausible. Smirnov himself inclined to favour the hypothesis of gas transfer (according to C. Fenner and N. Bowen) pointing out that it "represents a simple explanation for forward motions of ore-bearing fluids over the system of pores and fissures. A striving of gas for limitless expansion is well able to provide transport for ore-forming material" (*idem,* p. 152).

Fenner and Bowen in a scheme of gas transfer of ore matter developed in the 1930s supposed separation of ore-bearing gases from the granite and other magmas while they cool off. The same view of the role of gases in ore formation is also dominant currently. Accordingly, possible temperatures of ore-bearing gases depend on the temperatures of crystallization of granite magma, i.e. they are of 1000° C or lower. However, no compound is known to exist in a gaseous form under such temperatures and to be capable of transferring heavy metals. This provided grounds for a predominant recognition of hydrothermal theory, both by Smirnov in the mentioned work and, more widely, in the decades which followed.

The eduction model offers a new basis from which to treat the problem of gaseous (fluid) forms of the transfer of ore matter. The ore-bearing gases and fluids taking part in an eductional process are independent of the magma's crystallization. In all probability, magmas originate at rather high levels of the hanging wings of eduction zones, partially within the Earth's crust due to overall heating up and a drop in lithostatic pressure. Ore-bearing gases are liberated from the depth material of the out-thrusting lying wing at the mantle levels only. According to what was said above, initial temperatures of such gases may be 2000-2300° C which means that gases themselves can be of composition very different from that normally assumed. Under the above temperatures metal oxides and sulphides as well as some pure metals are able to exist in a gaseous state. Therefore, under the conditions of eduction, the possibility for gases to transfer ore material in the direction of the Earth's surface directly in compounds found in solid ore minerals should be recognised.

DEEP ORE-MAGMATIC CHANNELS

The question of links between the depth seismofocal zones and the upper levels of the Earth's crust has been repeatedly posed with respect to the above-seismofocal volcanic zones and can now be largely resolved. Geophysical research performed on the Kamchatka, Aegean region, Japan (Hedervari, 1973; Abdurakhmanov and Fedorchenko, 1976; Balesta, 1981) has established that beneath large volcanic central structures there are subvertical pipe-like formations going down to a depth of 200-250 km, i.e. to the level of surface of main movements at the lower boundaries of the seismofocal zones; their diameters are of a few tens of kilometres. These structure are detectable owing to increased absorption of seismic waves within them, in particular of transverse waves (which is assumed to be evidence of magmatic chambers), their aseismicity and other geophysical signs. They were given the name *magmatic columns* (Fedotov, 1974, 1975, 1976, 1979; Farberov, 1974; Balesta, 1981) and *translithospheric pipes* (Abdurakhmanov and Fedorchenko, 1976). The authors who described them think that they are channels which directly connect volcanic structures observed on the surface with the deep volumes of seismofocal zones. If revealed structures of such type are not considered to be exceptions (and this is hard to substantiate) one should, according to a wide distribution of such central volcanic structures, draw the conclusion that the entire lithosphere beneath volcanic zones (and belts) is pierced with vertical channels which start from the deep levels of seismofocal zones. These channels seem to be capable of acting as paths for direct transportation of matter from depth levels to the surface; participation of ore-forming compounds in this process can be proved by a known association of ores with some volcanic structures as well as by direct observations of depositions of mineral matter in the environs of active volcanoes. Such observations establish, in particular, participation of gaseous transfer of mineral matter. For example, after the 1975-1976 Tolbachik volcano eruption, S. F. Glavatskikh found in one of the newly-formed circular structures a vent of a high-temperature (about 500° C at outlet site) gaseous jet around which a crust of copper oxides, sulphates and chlorides (the main minerals are tenorite, melanotallite, dolerophanite in the form of crystal and globules and products of their alteration) 3-15 cm thick which was formed during a year and a half (Naboko and Glavatskikh, 1980).

It was stated above that the above-seismofocal volcanic zones (belts) constitute only a part of wider ore-magmatic belts which are disposed above seismofocal zones. The other part of these belts are an intrusive, or plutonic, zone, or, more accurately, a zone of intrusions of central type (also a perivolcanic zone). This zone is not inferior in width to the volcanic one and is often wider. It usually encompasses most mineral deposits of the ore-magmatic belt, among them the largest, and is characterized by distribution of intrusions whose age is about equal to that of volcanic covers in a volcanic zone. Typical features of many of these intrusions are comparatively small diameters (from a few tens to 6-8 km and smaller) and isometric, often regular round shape. They are frequently accompanied by concentric ring faults and in the present relief are observed as blocks of round section undergoing elevations or, less frequently, subsidences. According to the observations by the author made mostly in the North-East of Russia such intrusions should be identified into a special group of volcanic and plutonic structures based on their common morphological signs. Their composition changing from granites and granodiorites, often monzonites, to granosyenites and syenites, on the one hand, and to melanocratic diorites, gabbroides and, seldom, ultrabasites, on the other. Such intrusions are often identified as dome-like, which is hardly so, since a permanent reiteration of regular circle-like contours which bound many intrusions obviously eroded at different depths testifies to a predominantly pipe-like shape of intrusive-filled chambers and to their belonging to different-size stocks. If they were domes, eroded in rugged mountainous relief, they would have the accidental and intricate contours depending on the relief forms.

Such tecto-plutonic structures (as they are often surrounded by ring faults) of a central type, or, to be more exact, pipe-like chambers accommodating intrusions, distributed in the plutonic zones of above-seismofocal ore-magmatic belts, can, in the author's opinion, be compared with translithospheric pipes found beneath volcanic structures. Such supposition is prompted by the following circumstances:

1) disposition of the plutonic and volcanic zones above their common seismofocal zone;

2) an absence of a distinct border between the volcanic and plutonic zones; distribution of areas of volcanic activities in the form of irregular lobes and bays as well as isolated volcanic centres from a volcanic zone into the plutonic one over great distances;

3) permanently reiterated pipe-like forms of chambers accommodating central-type intrusions;

4) the similarity of shape and structure of tectono-plutonic formations in the plutonic zones with those of tectono-volcanic formation in volcanic zones (diameter sizes, ring fault boundaries, filling of central chambers by intrusions frequently of regular round section, etc.).

5) localization in connection with some tectono-plutonic structures, inside them, on their contacts or in their surroundings, of endogenic mineralization, morphologically similar in localization of mineralization in connection with tectono-volcanic structures.

If the above considerations are true, all the above-seismofocal ore-magmatic belt is encompassed with a system of depth ore-magmatic channels which make a special class of depth structures.

RELATIONSHIP BETWEEN MINERALIZATION AND INTRUSIONS

Distribution of intrusions and, more loosely, structures of a central type over the territory of mineralized belts, in particular of a Pacific type, was studied by many researchers, beginning from W. Lindgren, W. Emmons and others. E. Wisser considered isometric domes in sedimentary rocks, whose central parts often reveal intrusive formations, to be major ore-bearing structures in the west of the American continents. In some of his examples (Wisser, 1960) the dome-like structure of the enclosing rocks is hypothetical but central intrusive bodies do exist. Considerable attention has been given to the study of relations between central-type magmatic structures and mineralization by some Russian investigators. So, the works of M. A. Favorskaya and I. N. Tomson make wide use of the concept of isometric intrusive domes as a variety of ore-bearing `focal' structures. A special treatment of world distribution of central-type magmatic structures is contained in the book by Sveshnikova (1973). V. V. Solov'ev identify different-size radial and concentric fault structures of a central type whose central part is often occupied by intrusions; such structures (as well as `focal' structures) are, true enough, of different scales and do not always have a clear structural characteristic. Many specific essentially intrusive structures of central type of mainly funnel-like shape with steeply dipping bounds were described by Tugovik (1974; Mansurenkov (1979); Borisov and Borisova (1974), and others. They relate, to a large extent, to the volcanic zone but partly to the zone of central intrusions too.

The author of this book has attempted to resolve the question of the nature and signs of ore-localizing structures in the above-seismofocal mineralized belts. According to the view advanced herein, the main structures governing the arrival of ore matter from depth parts of seismofocal zones to the near-surface parts of the Earth's crust are depth ore-magmatic channels fixed by central volcanic structures or by central plutonic ones which are similar morphologically. The origin of such channels is most likely to be connected with eductional separation of large volumes of overheated gases and fluids from the outflowing mantle which find their way to the Earth's surface. Round-shaped intrusions observed in the Earth's crust occupy the upper parts of the channels and use them as chambers; magma intrusion seems not to have been a cause for formation of the chambers themselves; this is testified to, in particular, by the absence of tectonic deformations due to intrusion around the central type intrusions (excluding ring faults) which has been clearly established in some cases. The level of origin of most magmas which form intrusions of central type, judging from their predominantly acid composition, is the upper part of the lithosphere, while ore compounds enter in deep parts of the mantle in the vicinity of the lower border of seismofocal zones. Superimposing or close localization of magmatic bodies and ore mineralization depends on the fact that the same channels were used by both magma and ore-forming matter.

The works written by V. I. Smirnov and his collaborators in recent years pay special attention to the connection of endogenic mineralization with the Benioff zones which are considered by this author "as a deeply penetrating channel providing the arrival of depth endogenic matter into the upper zones" (Smirnov, 1974, p. 16). In his work devoted to energetic bases of post-magmatic mineralization, V. I. Smirnov makes calculations and arrives at a conclusion that the conductive heat of cooling of magmatic rock massifs is not enough to form large post-magmatic deposits, and their formation requires an advent of heat "in the form of powerful local convective flows acting for a long time, and arising at the expense of the Earth's internal heat" (*idem,* p. 16). Smirnov considers that conductors for such flows are highly permeable faults, and intrusive stocks, dyke clusters, volcano funnels and explosive pipes controlled by them. "Small intrusive massifs, and, for hydrothermal veins filling the very large massifs, when cooling off, cannot liberate enough heat energy for postmagmatic ore deposits to arise" (Smirnov, 1981, p. 10).

Some other authors also cite the data on heat delivery during formation of some intrusions which is more than the heat of the cooling magma. Such evidence supports the central thesis of this book that eduction is responsible for many features of the Earth's lithosphere.

REFERENCES

Abdurakhmanov, A. I., and Fedorchenko, V. I., 1976. On possible interpretation of spatial-and-temporal connections between seismic and volcanic events in the Kuril-Kamchatka zone (in Russian), in: *Volcanism in the Kuril-Kamchatka region and Sakhalin, Transactions of SakhKNII DVNTs*, iss. 48, Vladivostok, 6-16.

Alexandrov, A. A. 1978. *Overlapped and imbricate structures in the Koryak Highland* (in Russian), Nauka Publ. (MS).

Alexeev, E. S., 1982. Ophiolite complexes in the southern Koryak Highland (in Russian), *Geotektonika*, No. 4, 87-98.

Alvarez, W., 1972. Rotation of the Corsica-Sardinia microplate, *Nature-Phys. Sci.*, **235**, 103-105.

Andersen, T. V. 1991. Subduction and eduction of continental crust in the North Norway, *Terra Nova*, **3**, 308-310.

Artem'ev, M. Ye., 1966. *Isostatic gravity anomalies and some problems of their geological interpretation* (in Russian), Nauka Publ. (MS).

Artem'ev, M. Ye., 1975. *Isostasy of the USSR territory* (in Russian), Nauka Publ., (MS).

Artyushkov, Ye. V., 1980. *Geodynamics* (in Russian), Nauka Publ. (MS).

Aver'yanova, V. N., 1965. On general regularities in strees directions in earthquake foci of the Kuril-Kamchatka seismoactive region (in Russian), *Geotektonika*, No. 3, 85-98.

Aver'yanova, V. N. 1975. *Depth seismotectonics of island arcs* (in Russian), Nauka Publ. (MS).

Avigad, D., and Garfunkel, Z., 1991. Uplift and exhumation of high-pressure metamorphic terrains: the example of the Cycladic blueschist belt (Aegean Sea); *Tectonophysics*, **188**, 357-372.

Balakina, L. M., 1962. General regularities in directions of main tensions in the earthquake foci in the Pacific seismic belt (in Russian), *Izv. AN SSSR, Ser. Fizika Zemli*, No 11, 1471-1483.

Balakina, L. M., 1972. Tsunami and the mechanism of earthquakes foci in the North-West Pacific, in Tsunami waves (in Russian), *Transactions of SakhKNII DVNTs* , Issue 29, Yuzhno-Sakhalinsk, 48-72.

Balakina, L. M., 1979. Orientation of stresses in the earthquake foci in the Pacific seismic belt and the plate tectonics hypothesis (in Russian), *Izv. AN SSSR,. Ser. Fizika Zemli*, No 5, 16-30.

Balesta, S. T., 1981. *Earth's crust and magmatic chambers in the present-day volcanic provinces* (in Russian), Nauka Publ. (MS).

Barriga, F. J. A. S., Fyfe, W. S., Landefeld, L. A., Munha, J., and Ribera, A., 1992. Mantle eduction: tectonic fluidisation at depth, *Earth Sci. Rev.*, **32**, 123-129.

Bazhenov, M. L., and Burtman, V. S., 1994. Upper Creatceous paleomagnetic data from the Shikotan Islands, Kuril Arc—implications for plate kinematics, *Earth and Planetary Sci. Lett.*, **122**, 19-28.

Belousov, V. V., 1963. Concerning the Andes tectonics (in Russian), *Bull. MOIP, Otd. Geol.*, **XXXVIII** (2), 3-16.

Belyi, V. F., 1963. Concerning tectonic structure of the right side of the River Anadyr lower reaches (Laramides in the north-eastern Asia) (in Russian), in: *Data on geology and mineral deposits of north-eastern USSR*, Issue 16; Magadan, 106-126.

Belyi, V. F., 1977. *Stratigraphy and structures of the Okhotsk-Chukotka volcanogenic belt* (in Russian), Nauka Publ. (MS).

Bemmelen van, R. W., 1949. *The geology of Indonesia*, vol. 1, Government Printing Office, The Hague, The Netherlands.

Benioff, H., 1954. Orogenesis and deep crustal structure—additional evidences from seismology, *Bull. Geol. Soc. Am.*, **65**, 385-400.

Benioff, H., 1959. Circum-Pacific tectonics, *Publ. Dom. Obs.*., **20**, 395-402.

Benioff, H., 1962. Movements of major transcurrent faults, in: *Continental drift*, S. K Runcorn ed., Academic Press., NY-London, 103-134.

Bigi, G., Cosentino, D., Parotto, M., Sartori, R., and Scandone, P., eds. *Structural model of Italy and gravity map, scale 1:500,000*, 1992.; Quaderni de 'La Ricera Scientifica", No. 14, v. **3**, Florence.

Blinov, V. F. 1977. Development of the Pacific Ocean by the data of sedimentation and magnetic anomalies (in Russian), *Geologichesky Zhurnal*, **37,** No 2, 82-90.

Borisov, O. G., and Borisova, V. N., 1974. *Extrusions and related gaseous-hydrothermal processes* (in Russian), Nauka Publ., Novosibirsk.

Bragin, N. Yu., Grigor'ev, V. N., Krylov, K. A., and Sokolov, S. D., 1987. Triassic deposits in the western part of the Koryak Range (in Russian), *Geologia i geofizika*, No 2, 62-67.

Carey, S. W., 1959. *The tectonic approach to continental drift*, Univ. of Tasmania, Hobart.

Carey, S. W., 1975. The expanding Earth—an essay review, *Earth's Science Review*, **11** (2), 105-143.

Carey, S. W.,1976. *The Expanding Earth*, Elsevier, NY.

Chekhov, A. D., Alexandrov, A. A., Palandzhyan, S. A., and Petrov, A. N., 1980. Verkhnye-Khatyrsky serpentinite melange (central Koryak Highland) (in Russian), *Geologia i geofizika*, No 5, 51-59.

Chekhov, A. D.,1982. Tectonics of Talovka-Pekulnei zone (in Russian), in: *Studies on tectonics of the Koryak Highland*, Nauka Publ. (MS), 80-100.

Chudinov, Yu.V., 1962. Signs of superimposed heterochronous movements in the northeastern extremity of Tuva depression and in adjacent areas (in Russian), *Izv. AN SSSR, Ser. Geol.*, No 3, 88-99.

Chudinov, Yu.V., 1976. Expansion of the Earth as an alternative to the new global tectonics (in Russian), *Geotektonika*, No 4, 16-36 (English translation: 1977, *Geotectonics*, **10**, 240-251).

Chudinov, Yu. V.,1979. Direction of movements in the deep marginal oceanic zones, *Abstracts of Papers of XIV Pacific Science Congress (Khabarovsk), Committee B, Section* B1 (MS); 37-39.

Chudinov, Yu. V., 1980. Origin of the Alpine folding belt from the viewpoint of the Expanding Earth hypothesis (in Russian), in: *Tectonics of the Mediterranean belt*, Nauka Publ. (MS), 223-237.

Chudinov, Yu. V., 1981. The expanding Earth and tectonic movements: direction of movements in marginal oceanic zones (in Russian), *Geotektonika*, No 1, 19-37 (English translation: 1981, *Geotectonics*, **15**, 11-22).

Chudinov, Yu. V., 1984. Determination of Earth's sizes upon the paleomagnetic data (in Russian), in: *Problems of expansion and pulsations of the Earth*, Ye, Ye. Milanovsky, P. N. Kropotkin, Yu. M. Pushcharovsky, and L. I. Filatova, ed., Nauka Publ. (MS), 98-113.

Chudinov, Yu. V., 1985. *Geology of active oceanic margins and global tectonics* (in Russian), Nedra Publ. (MS).

Chudinov, Yu. V., 1992a. The development of lithosphere in conditions of Earth expansion accompanied by eduction, *Abstracts of Papers, 29th Intern. Geol. Congr.*, **1**, Kyoto, 46.

Chudinov, Yu. V., 1992b. The possibilities of exact determination of the existence of eduction or subduction, *Abstracts of Papers, 29th Intern. Geol. Congr.*, **1**, Kyoto, 150.

Coleman, R. G.,1971. Plate tectonics emplacement of upper mantle peridotites along continental edges, *J. Geophys. Res.*, **76**, 1212-1222.

Delonge, M. R., Wortel, M. Jr., and Spakman, W., 1994. Regional-scale tectonic evolution and the seismic velocity structure of the lithosphere and upper mantle—the Mediterranean region, *J. Geophys. Res.*, **99** (B 6), 2091-2108.

Dewey J. F. and Bird, J., 1970a. Plate tectonics and geosynclines, *Tectonophysics*, **10**, 625-638.

Dewey J. F. and Bird, J., 1970b. Mountain belts and the new global tectonics, *J. Geophys. Res.*, **75**. 2625-2647.

Dewey J. F., Pitman, W. C., Ryan, B. F., and Bonin, J., 1973. Plate tectonics and the evolution of the Alpine system, *Geol. Soc. Amer. Bull.*, **84**, 3137-3180.

Dickinson, W. R., 1971. Plate tectonic models of geosynclines, *Earth and Planetary Sci. Lett.*, **10**, 165-144.

Dietz R. S., 1961. Continent and ocean basin evolution by spreading of the sea floor, *Nature*, **190**, 854-857.

Dietz, R. S., 1962. Ocean-basin evolution by spreading, in: *Continental drift*, S. K. Runcorn, ed., Academic Press, NY-London, 289-298.

Dietz, R. S., 1968. Arthur Holmes: originator of spreading ocean floor hypothesis, reply; *J. Geophys. Res.*, **73**, 6576.

Dietz, R. S., and Holden, J. C., 1971. Pre-Mesozoic Oceanic Crust in the Eastern Indian Ocean (Wharton Basin)?, *Nature*, **229**, 309-312.

Dixon, J. M. and Farrar, E.,1978. Ridge subduction, eduction and the Neogene tectonics of the S. W. USA. *Geol. Soc. Am., Abstr. with Programs*, **10**, (7), 390.

Dixon, J. M. and Farrar, E., 1980. Ridge subduction, eduction and the Neogene tectonics of southwestern North America, *Tectonophysics*, **67**, 81-99.

Dmitriev, L. V., Sharaskin, A. Ya., and Garanin, O. V., 1976. Main features of ocean floor magmatism (in Russian), in: *Problems of Petrology*, Nauka Publ., 99-116.

Emery, K., 1965. Geology of the continental margin off Eastern United States, in: *Submarine geology and geophysics*, Whittard W.F. and Bradshaw, R. eds, Butterworth, London, 107-122.

Farberov, A. I., 1974. *Magmatic centers of volcanoes from seismological data*, (in Russian), Nauka Publ. (MS).

Fedotov, S. A., 1974. Relation of volcanoes to the Pacific focal layer, mechanism of magma rising and likely position of volcano-feeding mantle volumes (in Russian), in: *Geodynamics, magmas' formation and volcanism*, Petropavlovsk-Kamchatsky, 9-20.

Fedotov, S. A., 1975. Mechanism of magma ascent and deep feeding channels of island arc volcanoes, *Bull. Volcanology.*, **39** (2), 1-14.

Fedotov, S. A., 1976. Geophysical data on depth magmatic activity beneath Kamchatka and estimation of forces causing magma rising and volcanism (in Russian), *Izv. AN SSSR, Ser. Geol.*, No 4, 5-16.

Fedotov, S. A., 1979. On viscous heating of magma and lava flows. Diameters of asthenosphere magma columns and velocities of magma ascent and magma differentiation beneath island arcs volcanoes volcanism (in Russian), *Vulkanologiya i seismologiya*, No 1, 5-15.

Fyfe, W. S., Landefeld, L. A., Riberio, A., Barriga, F. J. A. S., and Munha, J., 1991. Mantle eduction, a second order process in plate tectonics (abstract of paper on conference of 1987), *Tectonophysics*, **191**, 419.

Geology of the USSR, **30,** 1970. North-East of the USSR, books 1 and 2 (in Russian), Nedra Publ. (MS).

Gerth, H., 1955. *Der geologische bau der Südamerikanischen Kordillere*. Gebrüder Borntraeger, Berlin—Nikolassee.

Gnibidenko, G. S., Bykova, T. G., Veselov, O. V., Vorob'yev, V. M., Kim Tschun Un, and Tarakanov, R. Z., 1980. *Tectonics of the deep-sea Kuril-Kamchatka Trench* (in Russian), Nauka Publ. (MS).

Gotsadze, O. D., Keilis-Borok, V. I., Kirillova, I. V., Kogan, S. D., Kukhtikova, L. N., and Sorsky, A. A., 1957. *Research in earthquake mechanism* (in Russian), V. I. Keilis-Borok ed., Transactions of the Geoph. Inst. Acad. Sc. USSR, iss. 40 (166), USSR Acad. Sci. Publ. (MS).

Grigor'ev, V. N., Sokolov, S. D., Krylov, K. A., Golozubov, V. V. and Pral'nikova, I. E., 1995. Geodynamic typification of the Triassic-Jurassic volcanic-siliceous complexes of the Kuyul's terrain (Koryak Range) (in Russian), *Geotektonika*, No 3, 59-69.

Hasegava, A., Umino, U., and Takagi, A., 1978. Double-planed deep seismic zone and upper-mantle structure in the Northeastern Japan arc, *Geoph. Journ. Roy. Astron. Soc.*, **54,** 281-292.

Hedervari, P., 1973. Concentration of earthquake energy in and around the Aegean volcanic belt, *Tectonophysics*, **19**, 369-381.

Heezen, B. C., 1962. The deep-sea floor, in: *Continental drift*, S. K Runcorn, ed., Academic Press., NY, London, 235-286.

Heezen, B. C., 1972. Inland and marginal seas, *Tectonophysics*, **13**, 293-308.

Hess, H. H., 1962. History of ocean basins, in: *Petrological Studies: a volume in honor of A. F. Buddington*, Geol. Soc. Am., NY, 599-620.

Hilgenberg, O. C., 1933. *Vom wachsenden Erdball*, Berlin.

Hilgenberg, O. C., 1966. Die Paläogeographie der expandierenden Erde vom Carbon bis zum Tertiar nach paläomagnetischen Messungen, *Geol. Rundschau*, **55**, H. 3, 878-924.

Hodgson, J. H., 1962. Movements in the Earth's crust as indicated by earthquakes, in: *Continental drift*, S. K Runcorn, ed., Academic Press., NY-London, 67-102.

Hodgson, J. H. and Stevens, A. E., 1964. Seismicity and earthquake mechanism, in: *Research in geophysics*, **2**: Solid Earth and Interface phenomena, Odishaw, H. ed., 27-59.

Honda, H., Masatsuka, A., and Ichikawa, M., 1967. On the mechanism of eatrhquakes and stresses producing them in Japan and its vicinity, *Geophys. mag*., **3**, 271-279.

Hoshino, M., 1976. *Eustacy in relation to orogenic stage*, Tokai Univ. Press, Tokyo.

Hoshino, M., 1983. *Marine geology* (in Japanese; translation into Russian: 1986, Nedra Publ., MS).

Initial Reports of the Deep Sea Drilling Project, 1970, **V**, Wash

Initial Reports of the Deep Sea Drilling Project, 1971a, **VI**, Wash

Initial Reports of the Deep Sea Drilling Project, 1971b, **VIII**, Wash

Initial Reports of the Deep Sea Drilling Project, 1972, **IX**, Wash.

Initial Reports of the Deep Sea Drilling Project, 1973a, **XIII**, Wash.

Initial Reports of the Deep Sea Drilling Project, 1973b, **XVI**, Wash
Initial Reports of the Deep Sea Drilling Project, 1973c, **XVII**, Wash
Initial Reports of the Deep Sea Drilling Project, 1973d, **XVIII**, Wash
Initial Reports of the Deep Sea Drilling Project, 1973e, **XIX**, Wash
Initial Reports of the Deep Sea Drilling Project, 1973f, **XX**, Wash
Initial Reports of the Deep Sea Drilling Project, 1974a, **XXII**, Wash
Initial Reports of the Deep Sea Drilling Project, 1974b, **XXVI**, Wash
Initial Reports of the Deep Sea Drilling Project, 1974c, **XXVII**, Wash
Initial Reports of the Deep Sea Drilling Project, 1975a, **XXXI**, Wash
Initial Reports of the Deep Sea Drilling Project, 1975b, **XXXII**, Wash
Initial Reports of the Deep Sea Drilling Project, 1976a, **XXXIII**, Wash
Initial Reports of the Deep Sea Drilling Project, 1976b, **XXXIV**, Wash
Initial Reports of the Deep Sea Drilling Project, 1978, **XLII A**, Wash
Initial Reports of the Deep Sea Drilling Project, 1980a, **LVIII**, Wash
Initial Reports of the Deep Sea Drilling Project, 1980b, **LIX**, Wash
Initial Reports of the Deep Sea Drilling Project, 1981a, **LX**, Wash
Initial Reports of the Deep Sea Drilling Project, 1981b, **LXII**, Wash
Initial Reports of the Deep Sea Drilling Project, 1981c, **LXIII**, Wash
Initial Reports of the Deep Sea Drilling Project, 1982a, **LXVI**, Wash
Initial Reports of the Deep Sea Drilling Project, 1982b, **LXVII**, Wash
Initial Reports of the Deep Sea Drilling Project, 1986a, **LXXXV**, Wash
Initial Reports of the Deep Sea Drilling Project, 1986b, **XCII**, Wash

Isacks, B., Oliver, J., and Sykes, L. R., 1968. Seismology and the new global tectonics, *J. Geophys. Res.*,**73**, 5855-5899.

Isacks, B., and Molnar, P., 1971. Distribution of stresses in the descending lithosphere from a global survey of focal-mechanism solutions of mantle earthquakes, *Rev. Geophys. and Space Phys.*, **9**, 103-174.

Kaigorodtsev, G.G., 1963. The nature of limestone cliffs in the River Vayega basin (in Russian), *Materials on geology and mineral deposits of north-eastern USSR*, Issue 16, Magadan, 310-315.

Karig, D. E., 1971. Origin and development of marginal basins in the Western Pacific, *J. Geophys. Res.*, **76**, 2542-2561.

Karig, D. E., 1977. Growth patterns on the upper trench slope, in: *Island arcs, deep sea trenches, and back-arc basins*, M. Talvani and W. C. Pitman, ed., AGU, Washington, 175-185.

Kawai, N., Ito, H., and Kume, S., 1961. Deformation of the Japanese Islands as inferred from rock magnetism, *Geophys. J. Roy. Astron. Soc. London,* **6**, (1), 124-130.

Keith, M. L., 1972. Ocean-floor convergence: a contrary view of global tectonics, *J. Geology*, **80**, 249-276.

Kidd, R. B., and Davies, T. A., 1978. Indian Ocean sediment distribution since the Late Jurassic, *Marine Geology,* **26**, 49-70.

Kononov, M. V., 1989. *Plate tectonics of northwestern Pacific* (in Russian), Nauka Publ. (MS).

Kosygin, Yu. A., 1982. Notes on geotectonic hypotheses. The expanding Earth (in Russian), *Tikhookeanskaya geologiya*, No 6, 118-122.

Krasny, L. I., and Fedynsky, V.V., eds., 1976, *Map of the seismicity of the Pacific mobile belt and the Pacific Ocean (1896-1968), scale 1:10,000,000*, Mingeo USSR, Leningrad.

Koziar, J., 1980. Expansion of ocean floor and its connection with hypothesis of Earth's expansion (in Polish). *Sprawozdania Wroclawskiego Towarzystwa Naukowiego,* **35 B**, 13-18.

Kropotkin, P. N., 1972. Character of tectonic processes in island arcs of the Far East and their age (in Russian), in: *Earth crust of islands arcs and Far-East seas: Results of investigations under International Geophysics Projects,* Nauka Publ. (MS), 51-68.

Kropotkin, P. N., 1980. Problems of Geodynamics (in Russian), in: *Tectonics investigations of the Geological Institute of Acad. Sci USSR*, Nauka Publ. (MS), 176-247.

Kropotkin, P. N. and Shakhvarstova, K. A., 1965. *Geological structure of the Pacific mobile belt* (in Russian), Transactions of Geological Institute of AN SSSR, Issue 134, Nauka Publ. (MS).

Kuzin, I. P., 1974. *Focal zone and structure of the upper mantle in the eastern Kamchatka* (in Russian), Nauka Publ. (MS).

Larin, V. N., 1980. *Hypothesis of the originally hydride Earth* (in Russian), Nedra Publ.. (MS).

Le Pichon X., 1968. Sea-floor spreading and continental drift, *J. Geophys. Res.*, **73**, 3661-3697.

Le Pichon, X., Francheteau, J., and Bonnin, J., 1973. *Plate tectonics*, Elsevier, Amsterdam.

Lisitsyn, A. P., 1974. *Sedimentation in oceans, quantitative distribution of sedimentary material* (in Russian), Nauka Publ. (MS).

Little, T. A., R. J. Holcombe. G. M. Gibson, G. M. Officer, P. B. Gans and M. O. McWilliams, 1992. Exhumation of late Paleozoic blue schists in Queensland, Australia, by extensional faulting, *Geology*, **20**, 231-235.

Lobkovsky, L. I., and Sorokhtin, O. G., 1979. Structure of underthrusting zones of lithospheric plates and origin of marginal seas (in Russian), in: *Geophysics of the oceans, v. 2,* Sorokhtin, O. G. ed., Nauka Publ. (MS), 183-194.

Lu, R. S., Pan, J. L., and Lee, T. C., 1981. Heat flow in the southwestern Okinawa Trough, *Earth and Planetary Sci. Lett.*, **55**, 299-310.

Ludwig, W. J., Ewing, J. I., Ewing, M., Murauchi, S., Don, N., Asano, S., Hotta, H., Nayakawa, M., Ichikawa, K., Noguchi, 1966. Sediments and structure of the Japan trench, *J. Geophys. Res.*, **71**, 2121-2137.

Lyustikh, Ye. N., 1975. Calculations for plate tectonics (in Russian), *Dokl. AN SSSR*, **221**, (4), 924-927.

Magnitskiy, V. A., 1965. *Interior structure and physics of the Earth* (in Russian), Nedra Publ., (MS).

Magnitskiy, V. A., and Artyushkov, Ye. V., 1978. Some general questions of Earth's dynamics (in Russian), in: *Tectonosphere of the Earth*, Nauka Publ. (MS), 487-521.

Mansurenkov Yu. P., 1979. *Volcanoes above intrusions* (in Russian), Nauka Publ. (MS).

Markov, M. S., G. Ye. Nekrasov., and S. A. Palandzhyan, 1982. Ophiolites and melanocratic foundation of the Koryak Highland (in Russian), in: *Studies on tectonics of the Koryak Highland*, Nauka Publ. (MS), 30-70.

Martignole, J., 1992. Exhumation of high-pressure metamorphic great terrains, a review; *Canad. J. Earth Sci.*, **29**, 37-45.

McElhinny, M. W., 1968. Northward drift of India—examination of recent paleomagnetic results, *Nature*, **217**, 342-344.

Meien, S. V., 1969. On the hypothesis of continental drift in Carboniferous and Permian from paleobotanic data (in Russian), *Geotektonika*, No 5, 3-16.

Menier, L., Roermund, H., and Lardeaux, L., 1991. Comparison of PT paths in allochtonous high-pressure metamorphic terrains from the Scandinavian Caledonides and the French Massif Central contrasted thermal structures during uplift, *Geol. Rundschau*, **80**, 333-348.

Milanovsky, Ye. Ye., 1983. Evolution and present state of the problems of the Earth's expansion and pulsations, *Intern. Geol. Review,* **25** (6), 621-640.

Milanovsky, Ye. Ye., 1995. Pulsation of the Earth (in Russian), *Geotektonika,* No. 5, 1995.

Moiseenko, F. S., 1975. Present-day concepts of Earth's crust (in Russian), *Leningrad State University Bulletin, Geology and Geography*, Issue 4, No 24, 5-16.

Moiseenko, F. S., 1981. *Grounds of depth geology* (in Russian), Nedra Publ., Leningrad.

Moiseenko, U. I., and Smyslov, A. A., 1986. *Temperature of the Earth's interiors* (in Russian), Nedra Publ., Leningrad.

Morelli, C., 1975. Geophysics of the Mediterranean, in: *Newsletter of the cooperative investigations in the Mediterranean*, Issue 7, Monaco, 29-111.

Morgan, W. J., 1968. Rises, trenches, great faults, and crustal blocks, *J. Geophys. Res.*, **73**, 1959-1982.

Murauchi, S., and Asanuma, T., 1977. *Seismic reflection profiles in the Western Pacific,* Univ. of Tokyo Press, Tokyo.

Naboko, S. I., and Glavatskikh, S. F., 1980. Exhalation copper mineralization during the post-eruptive stage of the Great Tolbachik fracture eruption (in Russian), *Vulkanologiya i Seismologiya*, No 4, 50-64.

Nekrasov, G. Ye., 1976. *Tectonics and magmatism of the Taigonos Peninsula and northwestern Kamchatka* (in Russian), Nauka Publ. (MS).

Okay, A. I. and Cedal, A. M., 1992. Evidence for intracontinental thrust-related exhumation of the ultra-high-pressure rocks in China, *Geology*, **20**, 411-414.

Otofuji, Y., and Matasuda, T., 1983. Paleomagnetic evidence for the clockwise rotation of Southwest Japan, *Earth and Planetary Sci. Lett.*, **62**, 349-359.

Packham, G, H., and Falvey, D. A., 1971. An hypothesis for the formation of marginal seas in the Western Pacific, *Tectonophysics*, **11**, 79-109.

Parfenov L. M., 1983. Continental margins, island arcs within Mesozoides of northeastern Asia and the kinematics of Mesozoic folding; paper 1, Mesozoides of Verkhoyansk-Chukotka area (in Russian), *Tichookeanskaya geologiya*, No 3, 3-26.

Peive, A. V., 1945. Deep-seated faults in geosynclinal provinces (in Russian), *Izv. AN SSSR*, Ser. Geol., No 5, 23-46.

Peive, A. V., 1947. *Tectonics of the North-Urals bauxite belt* (in Russian), MOIP Publ. (MS).

Peive, A. V., 1954. Asymmetry of deep-seated tectonic structures of the Urals-Tien Shan orogene and origin of its virgations (in Russian), *MOIP Bull., Otd. Geol.*, **XXII** (5), 105-124.

Peive, A. V., 1967. Faults and tectonic movements (in Russian), *Geotektonika*, No 5, 8-24.

Peive, A. V., 1969. Oceanic crust in the past geological time (in Russian), *Geotektonika*, No. 4, 5-23.

Peive, A. V., 1980. Deep-seated faults (in Russian), in: *Tectonics investigations of the Geological Institute of Acad. Sci USSR*, Nauka Publ. (MS), 95-122.

Pucher, R., Bannert, D., and Fromm, K., 1974. Paleomagnetism in Greece: indications for relative block movement, *Tectonophysics*, **22**, 31-40.

Puchkov, V. N., 1965. The problem of the movement of continents (in Russian), *Geotektonika*, No. 6, 95-110.

Prince, R. A., and Kulm, L. D., 1975. Crustal rupture and the initiation of imbricate thrusting in the Peru-Chile trench, *Geol. Soc. Am. Bull.*, **86**, 1639-1653.

Proceedings of the Ocean Drilling Program, Initial Reports, 1990a, **107**, College Station, TX.

Proceedings of the Ocean Drilling Program, Initial Reports, 1990b, **129**, College Station, TX.

Richter, C. F., 1958. *Elementary seismology*; W. F. Freeman , San Francisco.

Ringwood, A. E., 1975. *Composition and petrology of the Earth's mantle*, McGraw-Hill, NY.

Ritsema, A. R., 1964. Some reliable fault plane mechanisms of solutions, *Pure and Applied Geophys.*, **59**, 58-74.

Ritsema, A, R., and Veldkamp, J., 1960. Fault plane mechanisms of southeast Asian earthquakes, *Mededeverhandel, Ned. Meteool. Inst.*, 76.

Rogozhin, E. A., 1996. Focal mechanism of the Neftegorsk (Sakhalin) earthquake of May 27/28, 1995 (in Russian), *Geotektonika*, No 2, 45-53.

Rozhdestvensky V. S., 1975. Horizontal strike-slips in northeastern Sakhalin (in Russian), *Geotektonika*, No. 2, 85-97.

Rozhdestvensky V. S., 1982. Role of horizontal strike-slips in the structure of Sakhalin (in Russian), *Geotektonika*, No. 4, 99-111.

Ruppel, C., 1995. Extensional processes in continental lithosphere, *J. Geophys. Res.*, **100** (B12), 24187-24215.

Ruzhentsev S. V., Byalobzhesky, S. G.,Kazimirov, A. D., and Sokolov, S. D., 1979. Nappes and palinspastics of the Koryak Ridge (in Russian), in: *Tectonic development of the Earth's crust and faults*, Nauka Publ., (MS), 63-80.

Ryaboi, V. Z., 1979. *Structure of upper mantle of the USSR territory* (in Russian), Nedra Publ. (MS).

Sacks, I. S., 1977. Interrelationships between volcanism, seismicity and anelasticity in western South America, *Tectonophysics*, **37**, 131-139.

Sato, T., 1973. Several considerations on the deep sea plains, *Marine Sci. Monthly,* **5**, 55-59.

Sawkins, F. J., 1972. Sulfide ore deposits in relation to plate tectonics, *J. Geology*, **80**, 377-397.

Scalera G., 1991. Non-chaotic emplacements of arc-trenches zones in the Pacific Hemisphere, *Institute Nazionale di Geofisica Pubblicazione*, No 531, Roma.

Scalera G., 1995. Earth complexity vs. plate tectonic simplicity, in: *Frontiers of fundamental physics*, Barone, M., and Sellery, F., eds., Plenum, NY, London, 257-274.

Scholl, D. W., 1974. Sedimentary sequences in North Pacific Trenches, in: *The geology of continental margins*, Burk, A., and Drake, Ch., eds., Springer-Verlag, NY, Heidelberg, Berlin, 493-504.

Scholl, D. W., von Huene, R., and Ridlon, J. B., 1968. Spreading of the ocean floor: undeformed sediments in the Peru-Chile Trench, *Science*, **159**,869-871.

Scholl, D. W., Marlow, M. S., and Cooper, A. K., 1977. Sediment subduction and offscraping at Pacific margins, in: *Island arcs, deep sea trenches, and back-arc basins*, M. Talwani and W. C. Pitman eds., AGU,Washington, 199-208.

Sclater, J. G., and Fisher, R. L., 1974. Evolution of the east central Indian Ocean, with emphasis on the tectonic setting of the Ninetyeast Ridge, *Bull. Geol. Soc. Am.*, **85**, 683-702.

Seely D. R., Vail, P. R., and Walton, G. G., 1974. Trench slope model, in: *The geology of continental margins*, A. Burk and Ch. Drake eds., Springer-Verlag, NY, Heidelberg, Berlin, 249-260.

Seely, D. R., 1977. The significance of landward vergence and oblique structural trends on trench inner slopes, in: *Island arcs, deep sea trenches, and back-arc basins*, Talwani, M. and Pitman, W.C. eds, AGU, Washington, 187-198.

Seslavinsky K. B., 1970. Structure and development of the South-Anyui seam trough (Western Chukotka) (in Russian), *Geotektonika*, No 5, 56-68.

Shatsky N. S.,1946. Wegener's hypothesis and geosynclines (in Russian), *Izvestiya AN SSSR, Geol. Ser.*, No 4, 7-21.

Sheinmann, Yu. M., 1973, 1974. New global tectonics and reality, articles 1 and 2, (in Russian), *Bull. MOIP, Otd. Geol.*, **XLVIII** (5), 5-28, and **XLIX** (1), 5-26.

Sheridan, R. E., 1974. Atlantic continental margins of North America, *The geology of continental margins*, Burk, A. and Drake, Ch. eds, Springer-Verlag, NY, Heidelberg, Berlin, 391-407.

Shilo, N. A., Kulish, E. A., and Ignat'yev, A. B., 1979. Organization and metallogeny of volcano-tectonic structures in the Pacific surroundings (in Russian), *Dokl. AN SSSR*, **245** (6), 1471-1479.

Shilo, N. A., Merslyakov, V. M., Terekhov, M. I., and Tilman, S. M., 1973. Alazei-Oloi eugenosynclinal zone—a new element of Mesozoids in North-East of the USSR (in Russian), *Dokl. AN SSSR*, **210** (5), 961-964.

Shilo, N. A., Umitbayev, R. B., 1977. Mongol-Chukotka system of deep-seated faults—an ancient seismofocal zone (in Russian), *Geologiya i geofizika*, No 1, 158-154.

Shmidt, O. A., 1980. Bering Sea, Aleutian Island Arc, deepsea Aleutian Trench (in Russian), in: *Tectonics of continental margins in the Northwestern Pacific*, Nauka Publ. (MS).

Silantiev, S. A., Sokolov, S. D., and Polunin, G. V., 1994. New Data on the Composition of metamorphic units in the Talovka River Basin (Penzhin, Northeastern Russia) (in Russian), *Geotektonika*, No. 2, 82-90.

Sillitoe, R.H., 1972. Relation of metal provinces in Western America to subduction of oceanic lithosphere, *Geol. Soc. Am. Bull.*, **83**, 813—818.

Sillitoe, R.H., 1976. Andean mineralization: a model for the metallogeny of convergent plate margins, *Geol. Assoc. Canad.*, special paper, No. 14, 59—100.

Simbireva, I. G., Fedotov, S. A., and Feofilaktov, V. D., 1976. Heterogeneities of stress field of the Kuril-Kamchatka Arc according to seismological data (in Russian), *Geologiya i geofizika*, No 1, 70-85.

Smirnov S. S., 1945. Present-day state of theory of ore formation (in Russian), *Data on geology and mineral deposits of the north-eastern USSR*, iss.1, Magadan, 5-25 (also in: Selected works, 1955, AN SSSR Publ. MS).

Smirnov, V. I., 1974. Benioff zones and magmatogene ore formation (in Russian), *Geologiya rudnykh mestorozhd.*, **16** (1), 3-17.

Smirnov, V. I., 1981. Energetic base of postmagmatic ore formation (in Russian), *Geologiya rudnykh mestorozhd.*, **23**, (1), 5-17.

Smirnov, Ya. B., 1980. *Thermal field of the territory of the USSR (*explanatory note to the maps of the heat flow and abyssal temperatures, scale 1:20,000,000*)*, Main Adm. Geodesy and Cartography under the Council of Ministers USSR, (MS)

Smirnov, Ya. B., and Sugrobov, V. M., 1979-1980. Earth's heat flow in the Kuril-Kamchatka and Aleutian provinces, papers1, 2, and 3, (in Russian), *Vulkanologiya i seismologiya*, 1979: No 1, 16-31; 1980: No 1, 59-73, No 2, 3-18.

Smirnov, Ya. B., Sugrobov, V. M., and Galushkin, Yu. I., 1982. Heat flow in the the zone of joint of the Aleutian and Kuril-Kamchatka island arc system (in Russian), *Vulkanologiya i seismologiya*, No. 6, 96-115.

Sokolov, S. D. 1990. Active Pacific margins, the Koryak upland (in Russian), in: *Tectonic layering in the lithosphere and regional geological explorations*, Pushcharovsky, Yu. M. and Trifonov, V.G. eds, Nauka Publ. (MS), 225-239.

Sollogub, N. V., 1986. *Lithosphere of Ukraina*, Naukova Dumka Publ., Kiev.

Solov'eva I. A., 1981. On transverse faults in the mid-oceanic ridges (in Russian), *Geotektonika*, No. 6, 15-31.

Sorokhtin O. G., and Lobkovskiy, L. I., 1976. Mechanism of drawing-in of oceanic sediments into the underthrust zone of lithospheric plates (in Russian); *Izv. AN SSSR,. Ser. Fizika Zemli*, No 5, 3-10.

Spakman, W., Stein, S., van der Hilst, R., and Wortel, R., 1989. Resolution experiments for N. W. Pacific subduction zone tomography, *Geophys. Res. Lett.*, **16**, 1097-1100.

Stavsky, A. Ya., Chekhovich, V. D., Kononov, M. V., and Zonenshain, L. P., 1988. Palinspastic reconstructions of the Anadyr'—Koryak region from the standpoint of lithosphere plate tectonics, (in Russian), *Geotektonika*, No. 6, 32-42.

Sveshnikova, Ye. V., 1973. *Magmatic complexes of the central type*, Nedra Publ., (MS).

Sydora, L. J., Jones, F. W., and Lambert, R. St. J.., 1980. Model calculation of the thermal fields of subducting lithospheric slabs and partial melting, *Tectonophysics*, **62**, 233-241.

Synthesis of the Deep Drilling Results in the Indian ocean, 1978. Oceanography Series, 21, Elsevier, Amsterdam.

Tarakanov, R. Z., 1978. Model of environment structure in the focal zone and adjacent mantle of Kuril-Kamchatka and Japan region, in: *Earth's crust of Far-East island arcs and marginal seas*, Nauka Publ. (MS), 111-126.

Tarakanov, R. Z., 1981. Focal zones and their significance in island arc evolution, in: *Geology of Far-Eastern margin of Asia*, Vladivostok, 53-66.

Tectonics of Africa, translation into Russian of *Memoir sur la tectonique du continent Africain*, Choubert G., and Faura-Muret, A., eds., 1973. Mir Publ. (MS).

Thompson, G., Bryan, W. B., Frey, F. A., and Diskoy, J. S., Jr., 1978. Basalts and related rocs from deep-sea drilling sites in the Central and Eastern Indian Ocean, *Marine Geology*, **26**, 119-138.

Tilman S. M., Afitsky, A. I., and Chekhov, A. D.,1977. Comparative tectonics of Alazeya and Oloi zones (North-East of the USSR) and Kolyma massif problem (in Russian), *Geotektonika, No. 4*, 6-17.

Toksöz N. N., Minear, J. W., and Julian, B. R., 1971. Temperature field and geophysical effects of a downgoing slab, *J. Geophys. Res.*, **76**, 1113-1138.

Trümpy, 1960. Paleotectonic evolution of the central and western Alps, *Bull. Geol. Soc. Amer.*, **71**, 843-907.

Tugovik, G. I., 1974. *Explosions and ore-forming processes*, Nedra Publ., (MS).

Udintsev G. B., and Koreneva, B. V., 1980. Origin of aseismic ridges in the Eastern Indian Ocean (in Russian), in: *Geotektonika*, No. 6, 13-22.

Usov M. A. 1950. *Introduction in geology*, Acad. Sci. Kaz. SSR Publ., Alma-Ata.

Uyeda, S., 1974. Northwest Pacific trenche margins, in: *The geology of continental margins*, Burk, A., and Drake, Ch., eds., Springer-Verlag, NY, Heidelberg, Berlin, 473-491.

Vasilyev, B. L., and Udintsev, G. B., 1982. New data on the continental nature of the Obrutschev Rise (in Russian), *Dokl. AN SSSR*, **267**, (3), 694-697.

Vening Meinesz, F. A., 1940. The Earth's crust deformation in the East Indies, *Proc. Kon. Akad. v. Wetensch.*, Amsterdam, **48**, 278-293.

Vine, F. J., and D. H. Matthews, 1963. Magnetic anomalies over oceanic ridges, *Nature*, **199**, 947-949.

Visser, S. W., 1936. Some remarks on the deep-focus earthquakes in the International Seismological Summary, papers 1 and 2, *Gerl. Beiträge Geophys.*, **47,** 321-332, **48**, 254-267,

Vogt, P. R., Schneider, E. D., and Johnson, G. L., 1969. The crust and upper mantle beneath the sea, in: *The Earth's crust and upper mantle*, Hart, P. J., ed., Am. Geophys. Union, Geophys. monograph 13, Wash., 556.-617.

Wadati, K., 1928-1931. Shallow and deep earthquakes, *Geophys. Mag.*, (Tokyo): 1928, **1**, 162-202; 1929, **2**, 1-36; 1931, **4**, 231-235.

Wadati, K., 1935. On the activity of deep-focus earthquakes in the Japan Islands and nighbourhood, *Geophys. Mag.*, **8**, 305-326.

Ward, M. A., 1963. On detecting changes in the Earth's radius, *Geoph, Journ. Roy. Astr. Soc.*, **8** (2), 217-225.

Ward, M. A., 1966. Comments on detecting changes in the Earth's radius, *Geophys. J. Roy. Astr. Soc.*, **10**, (4), 445-447.

Wegener, A., 1915. *Die Entstehung der Kontinente und Ozeane*, F. Vieweg, Braunschweig.

Wisser, E., 1960. Relation of ore deposition to doming in the North American Cordillera, *Geol. Soc. Am. Memoir*, 77.

Yasui, M., D. Epp, K. Nagasaki, and T. Kishii, 1970. Terrestrial heat flow in the seas round the Nansei Shoto (Ryukyu Islands), *Tectonophysics*, **55**, 349-360.

Yoshii, T., 1979. A detailed cross-section of the deep seismic zone beneath northeastern Honshu, Japan. *Tectonophysics*, **55**, 349-360.

Zavaritsky, A. N., 1946. Some facts to be taken into account in tectonic constructions (in Russian), *Izvestiya AN SSSR, Geol. Ser.*, No 2, 3-12.

Zdorovenin, V. V., 1972. Thicknesses of sedimentation cover in the Pacific Ocean according to seismic data (in Russian), in: *Geophysics of the Pacific Ocean*, Nauka Publ., (MS).

Zelenov, K. K., 1972. *Volcanoes as sources of ore-forming components in sediments* (in Russian), Nauka Publ. (MS).

Zharkov, V. N. 1978. *Interior structure of the Earth and planets* (in Russian), Nauka Publ. (MS).

Zhitetsky, A. A., and Radzivill, V. Ya., 1964. On stratigraphy of the Paleozoic in the basin of River Orlovka, the right-hand tributary to the River Bolshoy Anyui (in Russian), *Data on geology and mineral deposits of north-western USSR*, Issue 17, Magadan, 82-87.

Zverev, S. M., 1964. Results from study of sediments in the Sea of Okhotsk and Kuril-Kamchatka zone of the Pacific (in Russian), in: *Structure of Earth's crust in the transitional zone from the Asian continent to the Pacific Ocean*, Nauka Publ. (MS)., 90-93.

INDEX

—F—

—G—

—H—

—I—

—K—

—M—

—N—

—O—

—V—

—W—